T0320351

Artificial Intelligence and Systems of the Earth

Artificial Intelligence and Systems of the Earth is a book about the potential and capabilities of artificial intelligence (AI) and machine learning (ML) for studying the Earth. It aims to serve as an eye-opener on new avenues of scientific research that can be enabled by AI/ML. This is not meant to be a 'how to' book but is written to answer the question 'what if'. It explains how these tools are currently being applied, and the new opportunities they have opened. Through many examples and application ideas from outside the Earth Sciences, the book discusses some of the most prevalent types of AI in current use, the future of AI hardware, and how AI/ML bring about change.

Features
- Provides accessible and compact coverage on the many uses AI in Earth Science.
- Covers AI, deep learning, and causal modeling concepts in an easy-to-understand language.
- Contains a chapter on generative AI and its specific strengths and challenges.
- Includes descriptions of computer hardware for AI and where it is headed.
- Offers a companion website with regularly updated content.

This book is an excellent resource for researchers, academics, graduate, and senior undergraduate students in Earth Science and Environmental Science and Engineering, who wish to learn how AI and ML can benefit them, its potential applications, and capabilities.

Artificial Intelligence and Systems of the Earth

Michel Speiser

CRC Press
Taylor & Francis Group
Boca Raton London New York

CRC Press is an imprint of the
Taylor & Francis Group, an **informa** business

Designed cover image: Cover art was part of an artist concept created by Tony DeVarco and Bonnie DeVarco for ICES. © 2009 Tony and Bonnie DeVarco, Bob Bishop and the International Centre for Earth Simulation. Some of the source imagery used to make this work is courtesy of NASA Earth Observatory.

First edition published 2024
by CRC Press
2385 NW Executive Center Drive, Suite 320, Boca Raton FL 33431

and by CRC Press
4 Park Square, Milton Park, Abingdon, Oxon, OX14 4RN

CRC Press is an imprint of Taylor & Francis Group, LLC

ISBN: 978-1-032-71050-1 (hbk)
ISBN: 978-1-032-71053-2 (pbk)
ISBN: 978-1-032-71052-5 (ebk)

DOI: 10.1201/9781032710525

Typeset in Latin Modern Roman
by KnowledgeWorks Global Ltd.

Access the companion website at https://book.aiml.earth

This work was inspired and funded by the International Centre for Earth Simulation (ICES) Foundation, Geneva, Switzerland, as core to its principal mission, and the open access publication of this book has been published with the support of the Swiss National Science Foundation.

To Laurence, Clémentine, Juliette, and Lily.

Contents

Preface   xi

Foreword   xiii

Acknowledgments   xv

About the author   xvii

1   Introduction   1

2   AI refresher   4

    2.1   Artificial intelligence 4

    2.2   Machine learning 5

    2.3   Training data 7

    2.4   Stochastic gradient descent (SGD) 7

    2.5   Overfitting 8

    2.6   Regularization 9

    2.7   Artificial neural network (ANN) 10

    2.8   Deep learning (DL) 10

    2.9   Dropout 12

    2.10   Convolutional neural network (CNN) 12

    2.11   Recurrent neural network (RNN) 15

    2.12   Transformer 15

    2.13   Reinforcement learning (RL) 16

    2.14   Generative model 17

    2.15   Diffusion model 17

    2.16   Transfer learning 18

    2.17   Causal model 18

3   Current and future applications of AI in Earth-related sciences   20

    3.1   Summarization and dimensionality reduction 20

3.2   Compiling datasets   23
3.3   Surrogate models   24
3.4   Model bias estimation   25
3.5   Computational stepping stones   26

4   AI and challenges in Earth-related sciences   28
4.1   Correlations/teleconnections   28
4.2   Cross-talk and weak signals   29
4.3   Non-linearity   30
4.4   Feedback loops   31
4.5   Phase changes   32
4.6   Chaos   33

5   AI hardware and quantum computing   35
5.1   Data and compute power   36
5.2   Hardware co-evolution   37
5.3   Training and inference   39
5.4   Quantum computing   40
5.5   How will AI and quantum computing shake hands? . . .   42

6   Why believe AI? The role of machine learning in science 44
6.1   Testability and complexity   44
6.2   The purpose of science   46
6.3   High-dimensional output, low-dimensional internals . . .   48
6.4   Data impedance mismatch and end-to-end DL   49

7   Generative AI   51
7.1   Transfer learning and fine-tuning   52
7.2   Unsupervised learning and generative models   53
7.3   Limitations   54
7.4   Implications for the Earth sciences   58
7.5   Outlook   59

8   Causal models: AI that asks 'why' and 'what if'   61
8.1   Causation vs correlation   61
8.2   Causal graphs   62
8.3   Causal inference   65
8.4   Assumptions and limitations   66
8.5   Causal discovery   67
8.6   Interactions with machine learning and deep learning . .   69

9 Conclusion **71**

Resources **73**
    Climate Change AI 73
    ClimateSet 73
    Earth on AWS 74
    Earth System Science Data 74
    Google Earth Engine 74
    Hugging Face 75
    Kaggle 75
    ML4Earth 75
    Pangeo 76
    Radiant MLHub 76
    Sentinel Hub 76
    SpaceML 77
    WeatherBench 2 77

Bibliography **79**

Index **93**

Preface

During my time as a data mining and machine learning researcher at the IBM Research laboratory in Rüschlikon (Zurich), I heard that Bob Bishop, formerly Chairman and CEO of Silicon Graphics, was visiting the lab to present his vision for an International Centre for Earth Simulation - ICES Foundation. Luckily, I was able to attend the lecture, and was very impressed by the ICES mission of building towards a holistic understanding of the Earth. The idea stuck in my mind for several years, and generated many discussions with Bob Bishop, which led to my joining the ICES Foundation as Chief Data Scientist in 2018. ICES seeks to add value through the horizontal integration of scientific disciplines, complementing the vertical specialized knowledge that is generated in universities. Building bridges between disciplines is thus key to our approach. An important step in this direction is to gather experts from multiple disciplines, and challenge them to think outside of their specializations, developing a common understanding and language, which is the purpose of the ICES Biennial Workshops, the seventh of which will be held in Geneva in October 2024. One thing that all disciplines have in common is that Artificial Intelligence has arrived fresh on the scene, offering a new and fast evolving set of powerful yet quirky tools. It became clear that keeping track of AI advances, and of their uses in science and technology, was an important and valuable service that I could contribute. This book and its companion website provide the reader with a compact presentation of AI fundamentals, as well as with an overview of present and possible future applications of AI, especially in the Earth sciences. We do not dwell on specific technical or implementation details, but rather aim at an increased understanding of the possibilities that AI opens up, as well as of some of its current limitations and deep pitfalls, and possible future directions. The website will be regularly updated as new developments arise, likely at an ever accelerating rate if the past few years are any indication.

Nyon, Switzerland, June 2024

Foreword

AI/ML has taken the world by storm and is causing disruption at all levels of society and science. Fortunately, it comes as we transition into the new era of data-intensive computing, and a corresponding retake of computing architecture itself. Thus, there is the wonderful convergence of possibilities – voluminous Earth Observation data, new systems hardware choices, and powerful new software prediction tools.

Like all eras of rapid innovation, we can expect to see a hybrid world of young and old blended systems as this global transition forges ahead. But most of all, we need clear thinking, clean definitions, yet adaptability and flexibility, as we lay down our roadmap of opportunity.

This book, *Artificial Intelligence and Systems of the Earth*, is a major contribution to the road ahead. It is a clear and precise summary of how these matters fit together at the fundamental level, and behaves as an over-the-horizon radar so to speak, of what's in store for both the professional and the public at large.

Combining developments of the recent past with breakthroughs of the present day, author Dr. Michel Speiser has built a bridge to our future – a book with countless chapters yet to be written, thus the parallel online version.

Emanating from work at the Geneva-based ICES Foundation, Dr. Speiser craftily helps us achieve that all elusive holistic view of Planet Earth, with its multitude of interacting and intersecting subsystems, co-existing and co-evolving within Nature at the micro, meso, and macro levels.

The *Symbiosis of Science, Society and Nature* is at the heart of our mission in the ICES Foundation, and we stand in awe of the immensity of the challenge.

Within such complexity, how exactly do we achieve that perfect balance and harmony on Earth for all life forms - that dynamic equilibrium and confidence in the predictability of our future? No matter how you answer such important questions, you will be amazed at the speed of progress brought about by the convergence of capabilities and insights already achieved by the AI/ML revolution that is now upon us. By means of this book, help is truly at hand.

Bob Bishop
President & Founder
ICES Foundation
https://icesfoundation.org

April 2024

Acknowledgments

I gratefully acknowledge the support of the ICES Foundation for this work, and in particular, I thank Bob Bishop for his insightful comments and suggestions on the evolving manuscript.

My deep appreciation goes to the reviewers of the book proposal or of parts of the manuscript, whose thoughtful feedback helped me improve the content: Ghassem Asrar, Jean-Philippe Fricker, Aleksander Molak, Jagadish Shukla and Olivier Verscheure, as well as an anonymous reviewer. Many thanks also to Laurence Bodinier and Florence Balthasar for proofreading.

Front cover: Cover art was part of an artist concept created by Tony DeVarco and Bonnie DeVarco for ICES.
© 2009, 2022 Tony and Bonnie DeVarco, Bob Bishop and the International Centre for Earth Simulation. Some of the source imagery used to make this work is courtesy of NASA Earth Observatory.

This work was inspired and funded by the International Centre for Earth Simulation (ICES) Foundation, Geneva, Switzerland, as core to its principal mission, and the open access publication of this book has been published with the support of the Swiss National Science Foundation.

About the author

Michel Speiser is the Chief Data Scientist for the International Centre for Earth Simulation (ICES) Foundation, an independent scientific non-profit foundation based in Geneva, Switzerland. Prior to this, he was a research staff member at IBM Research - Zurich. He holds a PhD from ETH Zurich, and Masters degrees in mathematical sciences (EPFL), complex adaptive systems (Chalmers University of Technology), and computer science and engineering (EPFL).

Michel maintains a web version of this book at https://book.aiml.earth

ORCID: 0000-0002-0032-2427

1

Introduction

Artificial intelligence, machine learning, and deep learning are terms that many scientists have seen appear and grow in the practice of their discipline, including those who work to understand the Earth and its many systems and processes. The reasons for this AI 'wave' are numerous, chief among them the fast progress in predictive ability achieved by deep learning since the early 2010s, which can deal with images, text, as well as numerical and other data types. This versatility also extends to the set of problem domains; AI/ML techniques are almost as widely applicable as computing itself, and wherever there is data to be learnt from, the odds are that some deep learning model is learning from it. We cover key concepts and definitions of these new technologies in Chapter 2. Much like the rest of this book, the coverage is not meant to be comprehensive. Rather, it aims to provide the reader with a sufficient vocabulary to navigate the bestiary of models in current use, and to get a glimpse of the avenues and opportunities that could open up as a result. The main objective of the book is not to answer 'how to', but 'what if'.

While broadly applicable, AI is antithetical to science in several respects. Unlike a computer model built from first principles, an AI model can be hard to interpret (often called a 'black box'), which makes it difficult to trust. It is also thoroughly unparsimonious in the number of parameters, going against long-established scientific and statistical practice in this aspect as well. Furthermore, it struggles to provide adequate uncertainty quantification of its results, which is usually a scientific requirement. In the face of such serious objections, why do scientists even consider it as a potential tool in the scientific toolbox? The answer is simply that the predictive capabilities of AI are so advanced, that dismissing it is hardly an acceptable option.

Consequently, computer scientists are working diligently to smooth out some of AI's rough edges outlined above, and exploring how these new tools can be applied in a sensible and productive manner. Machine

DOI: 10.1201/9781032710525-1

1

learning researchers in computer science have been proactive in seeking useful solutions to address climate-related issues [1], and the domain scientists themselves are equally energetically investigating questions pertaining to their specialized fields of Earth science, as evidenced by significant activity in the past few years. For instance, the US Department of Energy conducted wide-ranging workshops 2021–2022 on the topic of 'artificial intelligence for Earth system predictability'[1], to determine how AI could best be used to obtain a substantial improvement in the predictability of the Earth's processes [2]. A new journal of the American Meteorological Society, entitled 'AI for the Earth Systems', was launched and its first issue appeared at the start of 2022 [3]. Its chief editor, Amy McGovern, recently commented that in her observation, AI in general was becoming accepted by scientists outside of computer science, as a way to help augment their capabilities to do foundational science [4]. This sentiment is echoed by the US National Academies of Sciences, Engineering, and Medicine, which organized a workshop on the topic 'AI for Scientific Discovery' in October 2023. Many current scientific initiatives include a strong AI component, for instance the USMILE[2] project aims to produce ML-assisted understanding and modeling of the Earth system, to name just one project in Europe, and a lot of analogous activity is taking place in many other jurisdictions, too. We will explore some applications and challenges in Chapters 3 and 4.

In addition, AI and high-performance computing (HPC) are converging, because both require large amounts of computational resources, and scientific HPC simulation executions increasingly comprise AI workloads. Among several examples, we may cite the MAELSTROM[3] project, which aims to build HPC AI for weather and climate forecasts. Hardware is a key part of the equation, as AI increasingly requires computer architectures that are able to deal with efficient data movement, beyond mere number crunching. Therefore, it becomes relevant to understand the current state of hardware, and possible directions in which hardware will evolve in the future, to envision some possibilities that will open up in this new space. Chapter 5 looks into such hardware questions, including the possible contributions of quantum computing.

---

[1]https://www.ai4esp.org/
[2]https://www.usmile-erc.eu/
[3]https://www.maelstrom-eurohpc.eu/

To follow up on the computer hardware considerations, we reflect on some fundamental questions in Chapter 6. Under which circumstances can we trust AI, in what ways can it be a useful and reliable tool in the scientific process, and which scientific rules of the road need to be revised in consequence? AI will not replace the established pillars of science: theory, experiments, and simulation. Yet, it can provide support to these pillars, and possibly constitute a pillar of its own. We approach such questions by comparing and contrasting AI models with conventional computer models, which were similarly scrutinized at the end of the twentieth century, when their use in science became widespread.

Chapter 7 explores 'generative models', a new breed of deep learning models that is fast evolving in present times. The objective of generative AI is to learn rich internal representations of datasets, which then enable the generation of novel datapoints. The ability to generate high-quality text has already burst onto the mainstream scene in the form of ChatGPT, a so-called large language model that has been trained on a large fraction of all text ever written. Similarly, image generation tools that produce pictures based on the user's input text are becoming more capable and finding experimental usage, with video generation becoming the new frontier. Although generative AI is still nascent, it offers interesting capabilities for scientific research.

Finally, Chapter 8 covers a different branch of the AI tree, known as causal models. Quite unlike deep learning, causal models are predicated on delivering fully interpretable results. They are grounded in probability theory, combined with causal hypotheses and logic, and they enable so-called causal inference: computing effects from causes, and vice-versa. We describe the main ingredients in causal models and discuss where they have been demonstrated to provide useful insight into the causal structure of Earth systems, e.g. in analyzing the surface pressure and temperature anomalies in the Pacific Ocean.

2

AI refresher

This present chapter covers definitions and descriptions of some important concepts in artificial intelligence (AI), machine learning (ML), and deep learning (DL), with a focus on the latter of the three. It is best read sequentially, since later entries can refer to earlier ones. We assume some prior exposure to the central ideas and do not aim to provide a comprehensive introduction. For a self-contained treatment of these subjects, we refer readers to the following textbooks: Russel and Norvig [5] for AI, Bishop [6] for ML, and Fleuret [7] and Goodfellow et al. [8] for DL.

2.1 Artificial intelligence

A definition of artificial intelligence (AI) is challenging to come by. In fact, we have no clear, universally accepted definition of *intelligence* in general. Most dictionaries relate intelligence to the ability to learn or acquire knowledge, and/or to apply knowledge judiciously in order to achieve one's objectives, but there are many variations on this theme, and such definitions lean on other words which are themselves hard to define. However, assuming that we have some definition of intelligence, AI can be defined as *the intelligence exhibited by a machine*. That is, AI is often interpreted as a contrasting term, to describe a type of intelligence which differs from 'natural intelligence', which is intelligence as we perceive it in humans (or animals). This is a fuzzy notion, especially because natural intelligence has a tendency of being redefined, as machines progressively acquire new capabilities. For example, the author recalls attending a university lecture wherein it was claimed that the game of *Go* would likely never be solved by AI, as it would require true intelligence. A few years after AI's victory over the best human players (more detail in Section 2.13), the game of Go is already beginning to be referred

DOI: 10.1201/9781032710525-2   4

to as a relatively simple and low-dimensional game compared to newer frontiers, such as the real-time strategy game *Dota*[1]. Alan Turing, who contributed foundational ideas to computing and AI long before the latter became a practical reality, eschewed an explicit definition of AI. Instead, he proposed a *test* of intelligence; a machine could be deemed to be intelligent if humans were unable to tell it apart from a human, based on its written output.

In common parlance, the term 'AI' is most often used synonymously with its latest, most successful subfield, which at the moment is *deep learning*. In scholarly settings, AI designates an entire academic discipline, the term having been coined by mathematician John McCarthy, at a seminal workshop held at Dartmouth College in 1956. It refers to the branch of knowledge which studies the use of computers in applications beyond rote calculation and tabulation tasks. Over the years however, numerous technical approaches have been pursued in the quest to make computers intelligent, some of which are depicted in Figure 2.1. Many earlier AI efforts attempted to build on formal logic and automated reasoning, and are often referenced under the term of *Expert Systems*. Machine learning is another subfield of AI.

2.2   Machine learning

Machine learning (ML) is a subfield of AI, aiming to create systems which are able to *learn from data*. Typically, an ML model is trained by iterating over data, updating the model's parameters in order to gradually improve its accuracy on a given task. We distinguish between *supervised* and *unsupervised* learning. In supervised learning, the aim is to learn a mapping from input data to output data, where both are supplied to the algorithm. The desired outputs are referred to as *labels*. A popular example of a supervised learning task is, given a picture of

---

[1] Dota stands for 'Defense of the Ancients', which is a user-modified version of the online game Warcraft III: Reign of Chaos. The sequel, Dota 2, is among the most popular online games, and is played in 'e-sports' tournaments involving millions of US dollars in prize money, cf. https://en.wikipedia.org/wiki/Dota_2. It is deemed to be highly complex because 1) the player has a very large number of possible moves, 2) the game is real-time rather than turn-based, and 3) each player has only partial knowledge of the game state, rather than a view of the entire board.

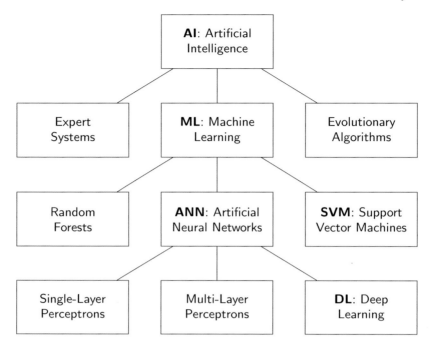

FIGURE 2.1 Hierarchical representation of AI and its sub-categories. This is non-exhaustive and is mainly intended to illustrate the place of deep learning within it.

either a cat or a dog, output the label 'cat' if the picture contains a cat, otherwise output the label 'dog'. Note that a label does not need to be text, it could also be a numerical value, or a picture, or of any other data type.

In unsupervised learning, no labels are available. The goal in unsupervised learning is often to produce a simplified description of the data. The canonical example is that of *clustering*: dividing the input datapoints into a certain number of groups, such that points within a group are more similar than points across groups. When the goal is to learn the internal structure of the data (such as next word prediction in text, as done by large language models), it is often called *self-supervised* learning because the labels are given by the data itself. An intermediate category of ML between supervised and unsupervised learning is referred to as *semi-supervised* learning, where we wish to learn a mapping from inputs to outputs; however, only a fraction of the inputs is labeled.

Mathematically, a supervised learning model can be described as:

$$\hat{y} = f(x, \theta) \tag{2.1}$$

The term *model* refers both to the function f and to the parameters θ. We provide input data x (e.g. the pixels of an image) and obtain the model's predicted output $\hat{y}$ (e.g. the label 'cat'). The model learns to produce good predictions, by being shown many pairs of examples (x, y) of input with true output. Often, the function f is fixed, and the training consists in finding values for θ such that predictions $\hat{y}$ are as close as possible to the given labels y, also called *ground truth*. For simple models, such as in linear or logistic regression, the best values for θ can be found using a single, closed-form expression. For more complex models, iterative algorithms are usually employed, which are generally not guaranteed to converge to an optimal solution.

2.3 Training data

In machine learning, the data is typically split into disjoint subsets, called *training set* and *test set*. Often, there is also a third subset, called *validation set*. The training set is used to fit the model, that is, the model is allowed to 'see' the training data in full. The optional validation set can be used to assess when to stop training the model, or to select one among several competing models. The final model's performance[2] is then estimated by applying it to the unseen test set.

A short note on the word *data*: although data is originally the plural of *datum* (which is defined as a piece of information), most of the ML literature uses data as a singular noun, and this book follows the same convention.

2.4 Stochastic gradient descent (SGD)

The most commonly used algorithm in machine learning is *stochastic gradient descent* (SGD). It proceeds by computing the gradient of the *loss*

---

[2]In machine learning, the *performance* of a model refers to its accuracy and precision in fulfilling the task, rather than to its speed or efficiency.

function (to be minimized) at each optimization step, not unlike Newton's method. The loss function measures[3] the difference between predicted outputs $\hat{y}$ and true labels y. The parameters are slightly modified in the direction of the gradient, in order to slightly improve the accuracy of the model. The 'descent' part of the name indicates that we follow the gradient downwards, continually decreasing the loss/error. The word 'stochastic' refers to the fact that we only use part of the data at each step of the algorithm, a small batch of randomly chosen data points. While gradient-based methods can be applied to a large class of models, in the case of neural networks, gradients are calculated using an efficient method called *backpropagation*, which implements the chain rule for differentiation. There exist several variations of SGD, called RMSProp [9], AdaGrad [10], Adam [11], etc. Gradients can be calculated automatically for nearly any given program, for example using the Autograd [12] reverse automatic differentiation system for the Python programming language.

2.5   Overfitting

In ML, we *fit* a model to data, using an algorithm such as SGD. In the supervised learning setting, this means that we search for parameter values θ such that the predicted outputs $\hat{y}$ are as close as possible to the true labels y given by the data. In this process, we can end up in the situation of *overfitting*, that is, we may find parameter values which are too closely molded to our particular dataset, which can happen if the function f has many degrees of freedom compared to the complexity of the data. This is illustrated in Figure 2.2, where three models (three different functions f) are fitted to the same data. In the leftmost panel, the model has too few degrees of freedom to capture the shape of the data, which is an *underfitting* situation. In the rightmost panel, we see the overfitting situation; the model is powerful enough to capture the shape of the data, but has in addition overzealously fitted small fluctuations that are likely due to sampling noise. The middle panel depicts a *proper* fit, which is the desired outcome in any ML application. Of course, in

---

[3]The loss function is usually a smooth version of the quantity that we wish to optimize. For example, in a classification task, the standard loss function is called cross-entropy, and it is used in place of the classification error rate because the gradient of the latter is uninformative (mostly flat with large jumps).

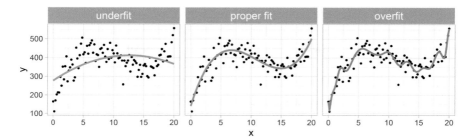

FIGURE 2.2 Illustration of the concept of overfitting. A given dataset of (x, y) points, shown as black dots, is fit with three models of increasing expressive power. Left: underfitting situation, the model is not expressive enough to capture the data's main shape. Middle: proper fit. Right: overfitting situation, the model attempts to capture minutiae of the dataset which are likely due to sampling noise.

general, a visual inspection of the fit is not feasible, especially when the dataset is very large and/or high-dimensional. How, then, does one detect when a model overfits? This can be achieved by observing how accurate the model is on a validation set, which is a part of the dataset that has been withheld from the training set, i.e. which was not used to train the model (cf. Section 2.3). When the model keeps getting more accurate on the training set, but starts getting increasingly worse on the validation set, it is a sign that the model has begun to overfit.

There are several techniques to avoid overfitting. The obvious solution would be to simply choose a function f that is appropriate for the complexity of the dataset. However, this is a very difficult thing to do in general, beyond simple situations. Several practical techniques are described in this document.

2.6 Regularization

A common technique to avoid overfitting is called *regularization*. It consists in adding a term to the loss function, to represent the degrees of freedom used by the model's parameters. Loosely speaking, the resulting loss function penalizes the model if it uses too many degrees of freedom, following the principle of Occam's razor[4]. This technique is broadly applicable because it makes no strong assumptions about the type of ML model being used; however, it can be difficult to select a good functional form and weighting of the regularization term.

2.7   Artificial neural network (ANN)

An *artificial neural network* (ANN), often just called *neural network*, is a very commonly used ML model. Its design is inspired by biological neurons, which are connected together, and get triggered if their inputs are sufficiently active. Mathematically, an artificial neuron is composed of just two things: 1) a linear combination of inputs, and 2) a non-linear function applied to this sum. This corresponds to the following expression:

$$output = activation \left(\theta_0 + \sum_{i=1}^{P} \theta_i \times input_i \right) \tag{2.2}$$

where the neuron receives a fixed number of inputs, and associates with each $input_i$ a parameter (or *weight*) θ_i. Note that a neuron's input can come directly from data (e.g. a pixel value) or from the output of another neuron. The term θ_0 is referred to as the *bias* of the neuron.

The activation function is a non-linear function. Historically, the sigmoid and tanh functions were predominantly used, but presently, piecewise linear functions such as the rectified linear unit (ReLU) [13], [14] are the most common, cf. Figure 2.3.

2.8   Deep learning (DL)

In a neural network, multiple neurons are connected together, the outputs of one neuron being fed as input into others. Neurons can be arranged in layers, such that the input data gets fed into the first layer, the outputs

---

[4]The possibly apocryphal Einstein quote, 'Everything must be made as simple as possible, but no simpler', describes the principle of Occam's razor, also known as the law of parsimony. This will be discussed in Chapter 6.

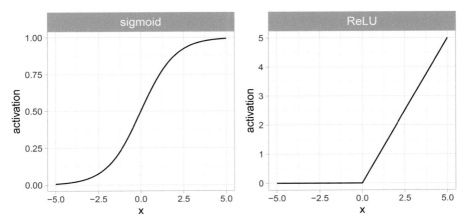

FIGURE 2.3 Non-linear activation functions commonly used in artificial neural networks. Left: Sigmoid function. Right: Rectified Linear Unit (*ReLU*).

of which get fed into the second layer, and so on, and the model output is provided by the final layer. The term *deep learning* (DL) refers to ANNs consisting of many layers. Such models were originally difficult to train efficiently, but the increased computational power of GPUs and ASICs, as well as algorithmic improvements, has made DL practical since the early 2010s; GPUs are extremely well suited to perform computations on matrices and tensors[5] (more on this in Chapter 5). The depth of the model is increased to enable the handling of complex data. For example, in the context of image data, the layers are able to represent concepts of increasing hierarchical aggregation. The first layer might capture basic concepts like edges and contrasts, the second layer might combine those concepts to capture simple shapes. These are in turn combined into more complex shapes, until finally in later layers, the neurons are able to detect a cat, or a dog. Such high-level 'features', such as cat or dog, emerge by themselves, simply by training a deep learning model. Previously, features were largely handcrafted in a labor-intensive process known as *feature engineering*, but deep learning has automated this and made it much more efficient.

---

[5]In this book, a tensor is an n-dimensional array of numbers. For example, a matrix is a two-dimensional tensor. Tensors are essential building blocks of deep learning, used to store both parameters/weights and data.

The *architecture* of a DL model refers to the types of neural layers used in the model, and to the topology in which they are interconnected. Below we list several common architectures, such as the CNN and the transformer.

2.9   Dropout

Dropout is an important overfitting avoidance technique in DL. It consists in disabling a random subset of the neurons during each training step. This forces the model to build in robustness and redundancy, and drastically reduces overfitting, even in models with an immense number of parameters (at the time of writing, the size of the largest models is crossing the trillion parameter mark [15], and the trend is towards ever bigger models).

2.10   Convolutional neural network (CNN)

One of the most common DL architectures is the *convolutional neural network*. The term refers to a specific way in which neurons in one layer are connected to neurons in the subsequent layer(s) of the network. The simplest architecture is called *fully connected*: the output of every neuron in layer k is connected to the input of every neuron in layer $k + 1$. The fully connected approach results in a very large number of parameters, and does not make use of the fact that pixels which are close to each other are more likely to be related. By contrast, in a CNN, a neuron in layer $k + 1$ receives as input only a small 'window' of pixels from layer k. In addition, the weights are the same for each window, which is referred to as *weight sharing*. These two architectures are compared in Figure 2.4. The convolutional concept can also be extended to higher dimensions; in fact, the most common application of CNNs has been to process 2D images since their inception [16], [17], cf. Figures 2.5 and 2.6. There are many variations of convolution, including the addition of padding around the input in order to obtain an equally sized output, or only applying the filter at each n-th position of the input (n is then called the *stride*), etc.

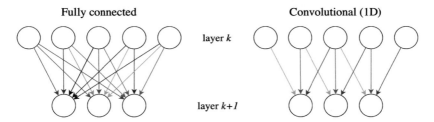

FIGURE 2.4 Two neural network architectures. Left: fully connected. Every neuron in layer k feeds into every neuron in layer $k+1$, and each connection has its own specific weight parameter, represented by the variety of colors. Right: convolutional (1D). Each neuron in layer $k+1$ receives input only from a small neighborhood of neurons in layer k. Moreover, the incoming set of weights is identical for each neuron in layer $k+1$.

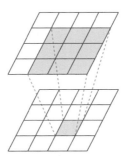

FIGURE 2.5 Simplified illustration of a single two-dimensional convolutional layer of depth one. The value of an element in the output tensor is computed based on a small neighborhood of elements in the input tensor, here using a window of size 3×3. A convolutional layer typically contains many filters and the input tensor can be of arbitrary depth, as shown in Figure 2.6.

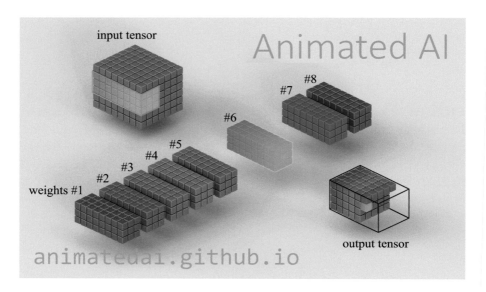

FIGURE 2.6 Illustration of the full computation performed by a 2D convolutional layer. This layer has eight sets of weights (also referred to as eight *filters*), numbered 1 to 8. The filters are size 3×3 and they have the same depth as the input tensor. Each filter is used in turn, in combination with the input tensor, to compute a slice of the output tensor, as indicated by the colors. This image is a frame taken from an animation, after filters 1 to 5 have already been applied, and filter 6 is being processed. At this moment, the filter is being multiplied by the highlighted section of the input tensor, and the nonlinearly transformed sum of this calculation is stored in the highlighted element of the output tensor. Once completed, the output tensor will have depth 8, the same as the number of filters. The full animation, along with other excellent deep learning animations created by Brad Klingensmith, can be viewed at https://animatedai.github.io.

2.11 Recurrent neural network (RNN)

Typically, neural networks are connected in a feed-forward topology, one layer being connected to the next, without 'loops' returning to a previous layer. Networks which contain such loops are called *recurrent neural networks* (RNN). Such networks are much more difficult to train than the feed-forward kind, since the loops can lead to numerical challenges in the learning process. Large gradients get even more amplified, while small gradients quickly go to zero, both of which impede a smooth gradient descent process. Despite this challenge, RNNs have seen significant research interest, because of their capability to handle variable-length inputs and outputs, as is the case for example in text translation from one language to another; sentences can be of any length, and the length of a French sentence is often not the same as that of its English equivalent. This class of learning problems is often referred to as *sequence-to-sequence* problems. recurrent neural networks using designs such as Long Short-Term Memory (LSTM) were the most successful ML models for these tasks for several years, but they have largely been displaced by alternative architectures such as the transformer models.

2.12 Transformer

The term *transformer* refers to a neural network architecture that has become highly prevalent in the past few years. Transformers were invented in the context of sequence-to-sequence problems such as text translation [18]. In particular, they are designed to overcome the limited training efficiency of RNNs (they can make much better use of GPUs than RNNs can). However, transformers are not limited to textual data, but are also applicable to image processing (vision transformers (ViT)) [19], and in fact they turn out to be strong competitors to the CNN models which have been dominating that space for a decade. The mechanism at the heart of transformers is called *attention*, for it mimics the human ability to temporarily focus one's attention to some specific thing while ignoring others. In particular, it allows the model to fetch the pieces of information that provide the most relevant context to the particular processing step being performed, even if that information is not in the immediate vicinity of each other. In a typical English sentence, one or a few words carry most of the important meaning, while the rest of the words act as support and are much more predictable. Since the real information content is concentrated in a few select locations, rather than uniformly spread out over the entire sentence, it makes sense to pay particular attention to those important locations, and the transformer integrates several such 'attention heads' in its design.

2.13 Reinforcement learning (RL)

Reinforcement learning (RL) is an area of machine learning where we wish to learn a function based on data that is only sparsely labeled. Typically, the objective is to learn an agent strategy involving a possibly long and complex sequence of actions, to achieve a certain outcome. For example, think of a game of chess, wherein one moves many pieces, in response to the moves of one's opponent, until ultimately arriving at the game's conclusion (win/loss/draw). Learning exactly which combination of actions – or avoidance of actions – led to a certain outcome, can be very difficult. A similar challenge arises when designing the control system of a robot with many degrees of freedom, to perform tasks in a complex environment. Reinforcement learning is the study of ML approaches in settings such as these, and has been in the popular spotlight since the company DeepMind used it in combination with deep learning, in a series of highly successful game-playing algorithms, such as AlphaGo and AlphaZero, which are able defeat the world's best human players in the game of Go. The word 'reinforcement' is borrowed from psychology, where positive reinforcement refers to the rewarding of good behavior, such as giving a treat to a dog for obeying a command. This same principle is applied to learning a good strategy.

RL has been applied to large language models (LLMs) in order to make them more conversational or otherwise appropriate. A model that was previously trained on an enormous body of text in an unsupervised fashion, is then fine-tuned using painstaking human annotations or scoring for some of its raw outputs. In this context, the technique is referred to as reinforcement learning from human feedback (RLHF) [20].

2.14   Generative model

A *generative model* is a statistical or ML model that approximates the joint probability distribution $P(x, y)$ of inputs and outputs (or simply the probability distribution $P(x)$ of inputs), rather than the conditional probability $P(y|x)$ of output given the input. In the latter case, the model is usually called a *discriminative model*. Once trained, a discriminative model is able to tell us whether there is a bird in a given input picture, whereas a generative model can allow us to sample from the input distribution, in other words, *to generate a (previously unseen) bird*. There are several approaches to creating generative models, including Generative Adversarial Networks (GANs), Variational Auto-Encoders (VAEs), diffusion models, etc. Generative models are a fast-evolving area of research and development, and Chapter 7 is dedicated to discussing their current state of the art and frontiers.

2.15   Diffusion model

A *diffusion model* is a generative model that is trained to progressively 'de-noise' a noisy image [21]. In this approach, it is very easy to obtain copious amounts of training data, by simply taking a high-quality image and gradually adding noise to it, cf. the sequence of images in Figure 2.7, from left to right. The model is trained to achieve the transformation in the opposite direction, i.e. from right to left. Ultimately, when the

FIGURE 2.7 Diffusion model illustration. An image is put through successive additions of noise (left to right). The ML model is trained to restore the image, which amounts to 'undoing' the noise addition. Such a model is then able to generate images from pure noise.

model becomes proficient at this task, we are able to inject a purely random signal, from which the model generates a plausible exemplar of the input distribution (e.g. a bird). This process can take place in pixel space, as in the illustration, or it can take place in a more abstract space. A popular approach of the latter idea is called *latent diffusion* [22], where the diffusion is performed in a low-dimensional latent space, allowing for more efficient learning.

2.16   Transfer learning

Transfer learning is a widely applied method in deep learning. It consists in repurposing a model for a different task than the one that it was trained for. This capability of DL came as a surprise in the field of machine learning, where beforehand, a model trained for one purpose was considered to be largely unusable for another, in most cases.

Training a large model from scratch can require extensive computational resources and infrastructure, and transfer learning enables the economizing of such resources, by leveraging prior training in a new context. Another important motivation for transfer learning is that there may not be a large labeled dataset available for learning the new task, and using a pretrained model can strongly reduce the data requirements. Typically, only a few layers of the deep neural network are significantly modified, while leaving the other layers largely unchanged. Adapting a model to a new task in this way is referred to as *fine-tuning* in the literature.

2.17   Causal model

All scientists are familiar with the adage, 'correlation does not imply causation'. We cannot establish a causal relationship between two variables merely based on an observed statistical association (e.g. correlation) between them. All of the machine learning techniques previously discussed in this chapter are based on statistical association. A *causal model* is a type of model that aims to capture causal relationships instead. They are beginning to be used to infer causation from time series data in

Earth system science [23]. Causal Models are a different branch of AI, and interested readers are referred to Pearl and Mackenzie [24] for a conceptual introduction, and to Peters et al. [25] for more technical depth. This topic will be expanded on in Chapter 8.

3

Current and future applications of AI in Earth-related sciences

In this chapter, we concern ourselves with some present day and possible future use of AI in scientific disciplines, mostly relating to the Earth system, including individual disciplines studying particular aspects of the planet, such as the atmosphere, the hydrosphere, the biosphere, the cryosphere, the lithosphere, etc., as well as Earth system science. However, we also consider technical sciences, which produce the myriad Earth-relevant datasets, obtained both through *in situ* observations and through remote sensing, as well as other disciplines from which inspiration can be gleaned concerning the productive use of AI.

3.1   Summarization and dimensionality reduction

Methods for summarization and dimensionality reduction have long been used in the sciences in order to reduce a complex data landscape to a more tractable level. This includes mathematical and statistical techniques such as singular value decomposition (SVD), and principal component analysis (PCA). AI and machine learning methods are increasingly being applied in the pursuit of this same objective. For example, machine learning has been used to produce a map of the ocean's physical regimes. A classical clustering algorithm known as k-means was applied to a dataset of physical quantities characterizing sea water dynamics, such as surface and bottom stress torque, bottom pressure torque, etc. This produced five clusters, corresponding to geographical areas with distinct physical regimes [26], [27]. Such a process allows researchers to distill a complex dataset down to a handful of patterns, which can each be studied in greater detail.

DOI: 10.1201/9781032710525-3

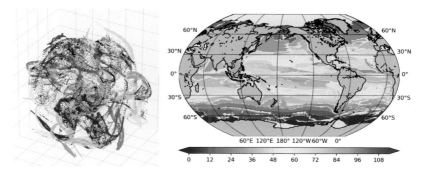

FIGURE 3.1 Marine eco-provinces mapped with machine learning. Left: t-SNE projection into a 3-dimensional space of the data set. Right: Spatial representation of the clusters obtained through DBSCAN (color indicates cluster ID) [30].

This use case is bound to continue and be built upon with newer ML techniques. For instance, a more recent algorithm based on deep learning can be applied towards the same end; the *t-SNE* [28] deep learning technique, in combination with a data mining algorithm called *DBSCAN*, was used to obtain a map of oceanic ecosystems [29], [30]. This two-step methodology is illustrated in Figure 3.1.

As discussed throughout this book, the traditionally preferred approach is to use a simple method whenever possible. In the context of identifying clusters in data, the clusters are then more *interpretable* than if a black box algorithm is used. However, in some cases, complex methods yield better results, so the question is perhaps akin to the military question of asking whether a map obtained from an enemy is preferable to no map at all. Researchers must judge which tradeoff between utility and interpretability is appropriate in each particular use case.

It should also be noted that AI can serve as *scaffolding*, to explore a dataset, and eventually be replaced by a more principled analysis. This could become an increasingly widespread modus operandi, as forays into explainable and interpretable AI techniques generate useful and portable tools. For example, consider the concept of saliency mapping, which allows the user of a convolutional neural network (one of the most prevalent deep learning variants in image processing, cf. Section 2.10) to inspect which areas of an image were predominant in determining the model's output. The most common saliency mapping technique at the time of writing is

FIGURE 3.2 Layer-wise relevance propagation (LRP) applied to the image of a rooster (left). The neural network correctly classified the image, and LRP (right) reveals which regions in the image were the primary influences to arrive at this classification; telltale features like the bird's comb and wattle are highlighted. Original photograph by Simon Waldherr (CC-BY), processed through http://heatmapping.org.

called layer-wise relevance propagation (LRP). The saliency map obtained by LRP in the case of a generic image classification example is shown in Figure 3.2. Although this technique is not fully robust (for example, it is not immune to adversarial attacks with images doctored specifically to deceive the model [31]), it can nevertheless provide some insight into the model's 'thinking', and thus provide the user with information upon which to judge its validity. LRP has been used in the context of investigating physical causes of climate change with ML. For instance, two Stanford researchers set about detecting the circulation patterns of extreme precipitation in images of 500 mb heights and of sea-level pressure, using as labels the observed precipitation data from stations, and PRISM precipitation data. They relied on LRP to get deeper insights into their model's mechanics [32].

3.2 Compiling datasets

The most successful application of deep learning has long been in the supervised learning setting (cf. Section 2.2), in particular for the image classification task, or variants thereof. The task amounts to determining the contents of an image. This is being applied to a plethora of research problems, such as tracking layers of ice in icesheets based on radar images [33], detecting the thawing of permafrost [34], identifying storm-signaling cloud formations such as the Above Anvil Cirrus Plume [35], or counting trees in the Sahel [36]. AI is indeed becoming a widespread tool among scientists, and scientific organizations are taking steps to integrate it into their processes. For example, NASA aims to increase the utilization of its image archive, by implementing 'search by image' functionality. This could enable researchers with a research question in mind, but initially having only a small data sample available, to assemble a larger dataset for study. Based on a relatively small set of exemplars supplied by the user, the system is designed to comb through the image archive, and extract image patches that bear a sufficient resemblance to the exemplars [37].

In each of these applications, AI acts as an identification mechanism for phenomena of interest, to be studied further by the scientists. Like an archaeologist's shovel, it helps us unearth finds. However, it may entice us to dig where the ground is softest, or allow us to mostly find objects that are buried shallowest. In other words, it can introduce *biases*. Understanding and handling such biases remains an important open problem. There is active research on using methods with increased explainability, such as Topological Data Analysis, applied e.g. for finding atmospheric rivers [38].

On a related note, machine learning methods for 'information fusion' are being explored very actively in the field of Earth observation, to merge data originating from a wide variety of sensors, stations, and model simulations [39]. Combining deep learning with process-based approaches for Earth system science has been recognized as an important challenge [40] and is an active area of research.

3.3   Surrogate models

Many processes of the Earth are very expensive to simulate, especially at fine resolutions. This is a crucial issue in climate modeling. Processes such as cloud formation and dynamics have significant effects on the warming of the tropical oceans; however, they occur at spatial scales that are orders of magnitude smaller than the width of a grid cell in even the most powerful climate models; on the largest supercomputers, the latter only operate on a grid with a spacing measured in kilometers, whereas the behavior of clouds requires going down to meter scale, or even much smaller if cloud nucleation physics is considered – see Figure 3.3.

Assuming that computers keep improving at historical rates (more on this in Chapter 5), we could see global climate models resolving low clouds by the 2060s [41]. In the meantime, climate modeling makes use of parametrizations for these subgrid processes. Rather than simulating physical first principles, a parametrization provides an approximation of the processes, fitted on empirical or simulated data. The same goes for numerical weather prediction, where relevant processes that are often parameterized include shallow and deep convection, latent and sensible

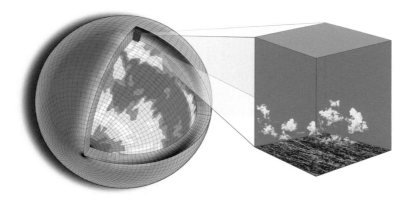

FIGURE 3.3 The grid of a global climate model is currently too coarse to model important processes (e.g. involving clouds) from first principles. Reprinted by permission from Springer Nature: Nature Climate Change, 'Climate goals and computing the future of clouds', Schneider et al. [41] © 2017.

heat flux, turbulent diffusion, wind waves, etc [42]. AI/ML models can be used in this way, and are increasingly being modified so as to incorporate physical constraints [43]. Accurately representing convective processes in climate models while retaining interpretability, is an active area of research, and some studies attempt to recover low dimensionality by using encoder-decoder architectures, with a parameter bottleneck in the mid-section of the model [44].

More generally, AI lends itself to surrogate modeling, when an outcome of interest cannot be easily measured or calculated. This can be useful in the case of parameter exploration – if a simulation is expensive to perform, parameters should be carefully selected, which can be assisted by a much cheaper surrogate model. Surrogate models are also seeing increased interest in forecasting applications. The AI4ESP workshop mentioned in Chapter 1 identified forecasting and understanding convective weather hazards as promising application areas, such as in the case of tornadoes [45], wind, hail, or lightning. Researchers from Nvidia, NERSC, and Caltech report an energy efficiency gain of approximately four orders of magnitude in their surrogate weather forecasting system FourCastNet, compared to the baseline weather model (although with lower skill) [46]. Their implementation relies on neural operators that perform much of the computational heavy lifting in the frequency domain, and make use of fast Fourier transforms (FFTs). Similarly, Google/DeepMind released a graph-based neural network (GNN) called GraphCast [47], and Huawei published its Pangu-Weather model [48] based on a three-dimensional transformer architecture, also showing remarkable predictive performance on mid-range weather forecasts.

3.4 Model bias estimation

In the above, we discussed surrogate modeling, wherein part of the model's work was delegated to AI. Another level at which AI/ML can be applied is *on top* of the physical model, post-processing or analyzing its outputs. A good example of this is an approach to bias estimation/correction devised by the European Centre for Medium-Range Weather Forecasts (ECMWF). Their numerical weather prediction system uses a data assimilation system called '4D-Var', which performs interpolation in space and time between a distribution of meteorological observations and the estimated model state. This data assimilation process results in some recognized biases, which can be estimated using a deep learning model that was pretrained on ERA5 reanalysis data[1] before being fine-tuned (cf. Section 2.16) on the computationally expensive integrated data assimilation system [49]. Their results showed a reduction of the temperature bias in the stratosphere by up to 50%. A related application of this idea can be found in studies of dynamical systems, under the name 'discrepancy modeling'. A nonlinear system is partly modeled with physical equations and constraints, and ML is used to deal with the residuals of this model with respect to observed data [50]. A comprehensive discrepancy modeling framework for learning missing physics and modeling systematic residuals, proposed in 2024, incorporates neural network implementations [51].

3.5   Computational stepping stones

AI methods are applied to an increasing number of domains, to come up with solutions that would take far too much time with conventional methods. For example, a DeepMind effort using a model named 'AlphaFold' has managed to produce a database of protein structures for over 200 million known proteins [52]. A protein's final structure is the result of *protein folding*, the physical process that translates a polypeptide chain into its stable three-dimensional form, which is notoriously expensive to simulate, and even more laborious to verify experimentally, such as through the use of X-ray crystallography. The structural predictions produced by AlphaFold may not be accurate in every case, but the database constitutes a rich resource for identifying interesting proteins, which can then be investigated through other methods. Possibly, an analogous approach could be used in the context of fluid dynamics, cloud formation, etc.

Machine learning is also being used in investigations as fundamental as determining the composition of subatomic particles. A recent study produced computational evidence for 'charm' quarks in protons, whereas the standard model only contains the 'up' and 'down' quark flavors, as per Figure 3.4. This was achieved by fitting parton distribution functions[2]

---

[1] https://www.ecmwf.int/en/forecasts/datasets/reanalysis-datasets/era5
[2] A parton distribution function gives the probability to find quarks and gluons in

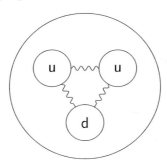

FIGURE 3.4 The established model for the proton contains two 'up' (u) quarks and one 'down' (d) quark, but no 'charm' quarks.

to thousands of experimental results, using neural networks [53]. In this case again, AI is a workhorse tool to propel scientific computing forward. Certain classes of problems encountered in the Earth systems may have similar characteristics to those cited above, and could therefore benefit from a similar approach.

For more in-depth coverage of deep learning applied to Earth Science, and in particular to remote sensing, climate science, and the geosciences, we refer the interested reader to the comprehensive book by Camps-Valls and colleagues [54].

---

a hadron (composite subatomic particle made of quarks, held together by the strong nuclear force) as a function of their percentage of the proton's momentum.

4

AI and challenges in Earth-related sciences

Many phenomena under study in the Earth-related sciences demonstrate aspects that we have come to describe as teleconnections, cross-talk, weak signals, non-linear dynamics, phase changes, and chaos. In this chapter, we outline how AI methods (in particular deep learning) are suitable for dealing with these issues.

4.1   Correlations/teleconnections

Deep learning is very effective at finding correlations in data, which can be harnessed to achieve high predictive power. Although correlations are easiest to discover when the spatial or temporal gap between the correlated signals is small, correlations can nevertheless be learned in the presence of large gaps as well, such as in the context of teleconnections, when predicting the regional water cycle based on the low-frequency climate modes of variability of El Niño Southern Oscillation (ENSO) [55]. While correlation does not equal causation, causation induces correlation, and detecting correlation can put scientists on the path of uncovering causal mechanisms. Increasingly, techniques are being developed which enable scientists to locate and characterize the underlying sources of correlation. Hence, while an ML model may not provide a causal explanation, it can be used to generate leads for investigating underlying physical causes. The correlation vs. causation question is examined in greater detail in Chapter 8. However, in the context of ENSO, saliency maps (Section 3.1) have been utilized to extract interpretable predictive signals from global sea surface temperature and to discover dependence structure that is relevant for quantitative prediction of river flows [56]. Machine learning has also been used to study relationships among teleconnections on a seasonal timescale, between the North Atlantic Oscillation, the Pacific

DOI: 10.1201/9781032710525-4

28

North American Oscillation, the West Pacific Oscillation, and the Arctic Oscillation [57].

4.2   Cross-talk and weak signals

When multiple phenomena are coupled in their underlying physical, electro-magnetic, or chemical make-up, through a form of energy transfer, these phenomena can be considered to act as a single system. We refer to this energy transfer as *cross-talk*, and it can occur in many engineered and natural settings. Two electrical circuits which are in close proximity can exhibit cross-talk from radiative effects, as can wave trains in the ocean that emanate from separate storms when they come close and interact. In such situations, a common approach is to consider each phenomenon as a subsystem. These subsystems are then assembled into a larger system by explicitly linking them together through a mechanistic or otherwise well-known scheme. By contrast, the typical DL approach is to consider the entire dataset from the viewpoint of a single model, letting the learning process itself figure out the dynamics of the full system. Alternatively, outputs for individual subsystems can be provided alongside the full data, so that the DL network can self-select any useful signals. Such approaches are proving very effective in engineering, in particular for the removal of crosstalk where it is typically undesired [58],[59]. In natural settings, the focus is typically not on crosstalk removal; however, it remains important to understand when and where it occurs. In biology, crosstalk refers to the intercommunication between different signaling pathways or cellular processes, involving the transfer of signals or molecules from one pathway to another, with a possible effect on the overall cellular response. The resolution of biological data having in some cases gone down to the single cell level, deep learning has been applied in the analysis of the resulting large datasets, with promising results across many topics, including in single-cell genomics and transcriptomics [60]. The above applications may provide inspiration to Earth Science practitioners and could be translated to some ES contexts, in particular in Earth system science.

In numerous scientific investigations, we are faced with *weak signals*, that is, signals which are largely or very nearly drowned out by noise, or are so sparse that they are difficult to measure. For instance, gravitational waves generated by black hole collisions, travelling across the cosmos, have only

recently become measurable thanks to LIGO detectors. In addition to
bespoke instruments, such weak signals may need special processing to
become evident, which can take the form of ML and DL[1] [62]. We may
also refer to weak signals in the context of scientific modeling, such as
when creating a mathematical model of a phenomenon where we have
discarded higher-order terms as being negligible, in order to make the
model more tractable and easier to study analytically. However, even if
those terms are small in comparison to the dominant ones, they may drive
a system behavior that turns out to be important, especially at different
spatiotemporal scales. In particular as the system approaches a tipping
point in its state of equilibrium, the interplay between small effects
can result in a non-negligible difference in outcome. The compounding
build-up of vorticity, turbulence and eddy currents from sub-micro level
origins into large-scale behavior patterns within the Earth's oceans and
atmosphere is a commonly recognized process. Neural networks are
showing promise for predicting ocean surface currents accurately, as
compared to physical simulation models [63].

4.3   Non-linearity

As was mentioned earlier, the default preference in science lies with
simpler models, all else being equal. Linear models are among the simplest,
relating a dependent variable with one or more independent variables
through a linear combination. One of the most ubiquitous statistical
models of this kind is the linear regression,

$$y_i = \beta_0 + \sum_{j=1}^{p} \beta_j x_{i,j} + \epsilon_i \tag{4.1}$$

where (x_i, y_i) is the i-th data point, x_i being a vector of size p. Furthermore, β_j is the j-th parameter of the model, and ϵ_i are the noise
terms, assumed to be independent identically distributed Gaussian

---

[1]In fact, deep learning can be so sensitive to weak signals that it is susceptible to
so-called 'adversarial attacks'. There is a subfield of neural networks that examines
how to trick a model into outputting the wrong answer, by purposely making minute
changes to the input data [61] (and how neural networks performing in vulnerable
environments could be made more robust).

random variables. This model is very popular, because its parameters lend themselves to a relatively straightforward interpretation, and because it has a single, closed form solution. Many natural phenomena involving numerous interacting and evolving processes on the other hand, cannot be modeled accurately using a linear model. The linear regression model can be generalized in various ways in order to deal with non-linear relationships; however, the aforementioned advantages decrease or disappear as models get more expressive, and choosing the correct type of model for a problem requires a high degree of expertise. In some such situations, deep learning can be a good alternative. Fitting a DL model to a large dataset is likely to require less domain knowledge and modeling proficiency than applying a tailored non-linear approach.

Differential equations constitute another very important scientific modeling tool, especially in the context of many Earth related disciplines. Here again, we need to 'switch gears' when non-linearity is introduced. Indeed, consider a dynamical system characterized by the equation,

$$\dot{x} = f(x, t) \tag{4.2}$$

where x is a vector, and the right-hand side is a vector field that depends on time t. If the function f is linear in x, the system is completely characterized by the eigenvalues of f. However, non-linearity in the function very often results in the absence of a closed-form solution, making the analysis, simulation or prediction concerning this system much more difficult. In many cases, in addition to being non-linear, f is effectively unknown, and machine learning can be a helpful tool to learn non-linear PDEs from data [64].

4.4 Feedback loops

In nature, we often encounter situations in which one phenomenon increases (or decreases) the frequency or intensity of another phenomenon. In many cases, this influence is bidirectional – the phenomena affect each other. We then speak of *feedback loops*. In particular, if each phenomenon has the effect of increasing the other in amplitude, the feedback is called *positive* or *self-reinforcing*. As an example, consider the arctic sea-ice melting from a warming ocean, decreasing the albedo of the planet, and causing the sea to absorb more energy from sunlight because of its darker color than if ice covered, thereby heating the upper ocean layers even higher, which in turn contributes to further sea-ice melting. Since feedback loops give rise to correlations, DL will be able to incorporate this signal for increased predictive power. However, a complex network of interacting positive and negative feedback loops may be difficult for a DL model to unravel, especially if it is not trained on a dataset which covers most possible states, or at least a sufficient selection of states such that interpolation between them leads to meaningful predictions. In the absence of sufficiently complete data, explicit modeling of domain knowledge is likely to be required, and in this regard the blending of neural networks with physics-informed partial differential equations can provide an answer [65].

4.5   Phase changes

The most familiar *phase changes* we encounter are those of water. When the temperature of water drops to zero degrees Celsius, it freezes; its *state of matter* changes from liquid to solid. Melting, vaporization and condensation are also phase changes – physical processes of transition between various states of matter, which occur when the pressure and temperature cross certain boundaries, as illustrated in Figure 4.1. More generally, we can think of a phase change as a qualitative shift in the basic structure and behavior of a system. Machine learning can be applied to identify when such shifts occur, which is especially useful when the physical parameters are not known in detail. Neural networks were used successfully for the classification of phase changes and states of matter in highly intricate settings, such as in quantum-mechanical systems [66]. Neural networks were also applied in combination with atomistic simulations and first-principles physics to generate phase diagrams for materials far from equilibrium [67]. Specifically, deep learning was used to learn the Gibbs free energy, and phase boundaries were determined using support vector machines (SVM). The obtained 'metastable' phase diagrams allowed the identification of relative stability and synthesizability of materials, and the phase predictions were experimentally confirmed in the case of carbon as a prototypical system. Phase diagrams are also of interest at much

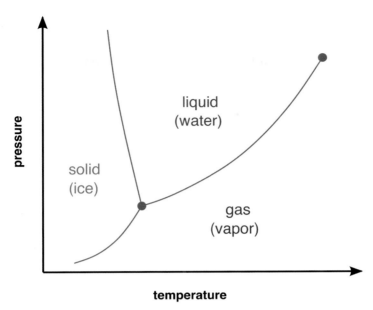

FIGURE 4.1 Phase diagram for water. When the system's state crosses a boundary (solid green line), a phase change occurs.

larger scales, such as in the context of the Earth's climate. Consider for example the schematic phase diagram shown in Figure 2 of *Lessons on Climate Sensitivity from Past Climate Changes* [68], plotting the planet's global mean surface temperature versus the atmospheric carbon dioxide concentration, featuring two disjoint branches: a 'cold' branch for a climate with polar ice sheets, and a 'warm' branch for a climate without them. Potentially, AI could assist in deriving such diagrams, but in more detail.

4.6 Chaos

Many phenomena in nature exhibit *chaotic* behavior, making them very hard to predict. In a chaotic system, the outcome is highly sensitive to initial conditions, a property that is often referred to as the 'butterfly effect'. A tiny change at the start of the process can lead to dramatically different final results. There is evidence that machine learning can be used

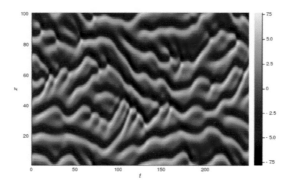

FIGURE 4.2 Plot of a simulation run of the Kuramoto-Sivashinsky flame equation. For each timestep on the horizontal axis, the flame front is described as a vertical strip of colors. Image credit: Eviatar Bach (Creative Commons CC0 1.0) using source code by Jonas Isensee.

to improve predictability even in such seemingly hopeless cases, as was demonstrated in the context of the Kuramoto-Sivashinsky equation, also called the 'flame equation' because it models the diffusive instabilities in a laminar flame front, a simulation of which is shown in Figure 4.2. A neural network model was trained to forecast the evolution of the system, without the model having access to the equation itself. The research team were able to achieve accurate predictions much further into the future than was previously thought possible [69], [70].

5

AI hardware and quantum computing

Computer hardware is the substrate of AI, just as the biological brain is thought to be the substrate of natural intelligence. Consequently, hardware is a key factor determining the abilities and performance of AI. According to one school of thought, AI is best pursued by imitating what we observe in biology. Such approaches could include commercial off-the-shelf (COTS) hardware, or may benefit from bespoke hardware. We are currently heading down the path of hardware specialization, although the wave of deep learning was launched on COTS hardware, which serendipitously matched its computational needs to a tee.

Indeed, the revolutionary deep learning model 'AlexNet' [71] was a CNN-based neural network (cf. Section 2.10), essentially a scaled-up version of its precursor model proposed some 25 years earlier [72]. AlexNet was implemented to run on Graphics Processing Units (GPUs), which were not initially germane to AI at all. GPUs were specifically designed to process and display graphical data, especially in 3D for the computer gaming sector. They had been steadily growing in performance for two decades, fueled by the public's voracious demand for high-powered gaming experiences. At their core, GPUs are built to perform matrix multiplications at high speeds. As it happens, this is precisely the workhorse operation underlying all of deep learning. Consequently, AI quickly added itself to the list of GPU customers, alongside gamers and more recently cryptocurrency miners, and the AI use case has been growing ever since, gaining official support from the major manufacturers. Many of today's largest supercomputers are equipped with hundreds or thousands of GPUs, and deep learning has become a staple scientific workload running on these machines.

DOI: 10.1201/9781032710525-5   35

5.1   Data and compute power

The concept of artificial neural networks, as well as the main algorithm for training them (cf. Section 2.4), has been around in some form since at least the 1960s[1]. However, they remained impractical until fairly recently, because two chief ingredients were not available in sufficient quantity: data and compute power.

In order to train a deep learning model from scratch, one typically requires a large dataset. The public availability of large datasets is a relatively recent phenomenon, and building such a dataset to get started was a significant hurdle which would have made this avenue of research impractical for most researchers. In fact, the AlexNet breakthrough was achieved in the form of an entry to the ImageNet Large-Scale Visual Recognition Challenge (ILSVRC) competition, using a subset of one of the first publicly available large collections of annotated images – over 1 million images, ranging over 1000 image categories [71].

The second ingredient whose availability to the average researcher lagged far behind the invention of neural networks is compute power, although high-end computers were available to the defense sector. As was previously discussed, stochastic gradient descent is an iterative algorithm, repeatedly cycling over the dataset's entries, slightly improving the model at every pass. This incremental training process is highly compute and memory intensive, and training a model of sufficient size to obtain interesting results would have been beyond the means of most major research institutes, let alone that of individual researchers. Interestingly, the research on neural networks had lain dormant for so many years, that the ImageNet winning model was trained on consumer grade hardware, rather than on a supercomputer.

In a prescient piece entitled 'The unreasonable effectiveness of data' [74], prominent Google researchers shared their observation that large volumes of data, fed into a simple (but large) model, often leads to better outcomes than spending one's efforts on building a more intricate and sophisticated model. They came to this conclusion while working on huge swathes of textual data, but it turns out to be applicable to image data

---

[1]The exact attribution of the invention of backpropagation is not without debate, but can be argued to date back to 1970 [73].

as well, and will surely extend to further data modalities. Prominent AI researcher Richard Sutton reached a similar conclusion and published the 'bitter lesson' [75] that he learned over the course of many decades, that computation trumps cleverness, because simple models are easier to *scale* in the long run. Kaplan et al [76] provided a striking illustration of this phenomenon in the context of large language models. They showed smooth relationships between the test loss (i.e. the quality of the learned model) and the amount of computation put into the training, the dataset size in number of words/tokens, and the size of the model in number of parameters. Based on those laws, it even became fairly *predictable* how well a model would perform, given a certain compute budget, a training set of a certain size, and a certain number of parameters. Such predictable benefits have spurred huge investments in AI hardware, mainly in the form of GPUs, as well as into custom AI hardware architectures. The increased investment is highly visible in Figure 5.1, which shows a steep uptick in computational resources spent on training state-of-the-art ML models. Where the trend was slightly outpacing Moore's law up until the early 2010s, it has grown much more rapidly since then (Moore's law is described in Section 5.4).

5.2   Hardware co-evolution

The company Nvidia oriented itself to AI as an important application early on, releasing their CUDA Deep Neural Network library (cuDNN) of deep learning primitives in 2014, allowing AI developers to make optimal use of their GPUs, and feeding the deep learning wave [78]. The rapid successes of DL across a wide variety of tasks and domains have turned from individual snowballs into a global avalanche, AI methods now fast becoming mainstream tools, and accordingly requisitioning a considerable share of the world's computational resources. In fact, DL workloads are now so widespread that they have begun to influence and shape the development of computer hardware itself. During the Transforming AI panel at 2024 Nvidia GTC conference, the authors of the transformer neural network architecture argued that the history of deep learning has been to 'build an AI model that's the shape of a GPU, and now the shape of a supercomputer', to which Nvidia's CEO Jensen Huang replied, 'we're building the supercomputer to the shape of the model'. Chips, circuits

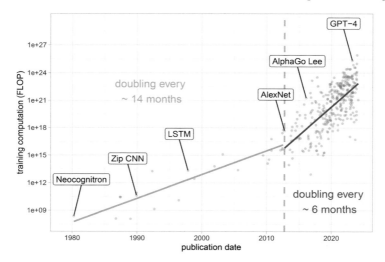

FIGURE 5.1 Floating-point operations (FLOP) necessary to train machine learning models, as a function of time. Each point represents a notable ML system according to the Epoch database [77]. The vertical dashed line splits the timeline into two segments, before and after the ImageNet 2012 competition. A regression line is displayed for each segment. The rate of increase in the computational requirements for model training has more than doubled since the advent of deep learning.

and systems are designed with AI use cases in mind, making AI a first class citizen in hardware design considerations, where a decade prior it was riding on the coattails of computer games. Today, entire datacenters are built for the sole purpose of making AI more efficient.

Google was a trailblazer in this new industry, stating in early 2016 that their Tensor Processing Unit (TPU) had been in use in their datacenters for over one year. The TPU was specifically designed to support their TensorFlow software framework for training and running deep learning models. This was their initial design; the fifth generation of TPU, which was announced in 2023, is orders of magnitude more capable. A host of companies has joined the fray to produce machines that are highly optimized for typical deep learning workloads. A key difficulty to overcome in this respect is the *scaling* of ML training – how to train ever bigger models?

Simply splitting the model training across multiple computers, linked together through a network, introduces inefficiencies. Such a system is bottlenecked by bandwidth and latency, meaning that the heaps of data can't make their way to the processing elements quickly enough. This situation has led the company Cerebras to design for maximizing the amount of computation that can possibly be achieved on a single semiconductor device. Indeed, their revolutionary concept is to use an entire silicon wafer[2] per unit, rather than follow the industry standard method of splitting the wafer into hundreds or thousands of chips. The processing elements on the wafer have the considerable advantage of being directly connected to each other via sub-micron copper wires, produced by extremely accurate lithographic processes. In the traditional approach, transferring data between processing elements frequently means moving data off one chip and onto another one, via a comparatively much bulkier interconnect using up space and energy, as well as leading to additional serialization/deserialization and caching components, whereas the Cerebras unit is able to shuffle the data around the wafer on shorter distances and using a flat memory hierarchy, which results in compounding efficiency gains. In addition, it integrates optimizations in the presence of sparsity of neural networks (many weights or activations are equal to zero, resulting in unnecessary computations), which can otherwise be remedied by additional (sometimes costly) procedures such as pruning [79]. Indeed, sparsity is an aspect that can crop up in many areas of AI/ML, modeling, and datasets. Efficiently handling this sparsity, in particular by avoiding unnecessary computation on zero-valued or null data, is likely to grow in importance as data sizes reach new heights, and can be tackled through special treatment at the hardware level [80].

5.3 Training and inference

Thus far in this chapter, we have mainly discussed the *training* phase of deep learning. This refers to the stage of creating a model, and teaching it to perform well on a given task, such as detecting a pedestrian in an image. The training phase is highly demanding in computational resources, and is typically done in a centralized setting where resources may be more abundant.

---

[2]Pure silicon for use in the semiconductor industry is usually prepared in a cylindrical ingot called *boule*, and a round slice of this ingot is referred to as a *wafer*.

Conversely, the *inference* phase of DL refers to the actual *use* of the model, once the training of its parameters is complete, and it is deployed in an operational setting. In certain cases, inference also takes place in a centralized setting. A good example of this is the processing of billions of queries to a large language model (LLM) such as ChatGPT. In many cases, however, inference is done in a decentralized fashion, close to the source of the input data. For instance, the pedestrian-detection model mentioned above could be installed as a component in a car's driver-assistance system. This is referred to as *edge AI*, because it operates somewhere at the edge of the internet, rather than at its center.

A central training facility requires hefty amounts of compute power, memory and bandwidth, whereas an inference/edge AI system typically performs much fewer calculations. However, it needs to be sufficiently fast to reliably produce outputs for its intended purpose, and has to run on a fairly small energy budget, so as to avoid draining the battery of the edge device (mobile phone, electric car, satellite, etc). Because of these key differences in requirements, the inference phase has also spawned a significant number of IoT designs, both from established manufacturers and from startup companies.

5.4   Quantum computing

Much like AI, Quantum Computing has made many headlines in recent years, and the two are similar in that both are technologies with tremendous potential and promise, although they are still undergoing fundamental developments.

Quantum computing aims to harness the laws of quantum mechanics to solve problems which are intractable for classical computers, in the sense that they would require far too much time and/or energy to solve. The key ingredient in quantum computing is the quantum *entanglement* of its memory elements[3], which allows them to work together as a single entity, if it can be maintained for a sufficient amount of time. Roughly speaking, this means that with every additional element, a quantum computer doubles the complexity that it can handle; its capability grows

---

[3]In the most common formulation of quantum computing, the memory element is called a *qubit* (short for *quantum bit*).

exponentially as we add computing elements. This is in stark contrast with classical computers, where adding a computing element (say, an additional CPU) results in a roughly linear[4] increase in capabilities.

This exponential possibility is what generates so much enthusiasm for the pursuit of quantum computing. The semiconductor industry itself has been accustomed to rapid exponential growth since its beginnings, the number of transistors in a dense integrated circuit having doubled approximately every two years for more than half a century, a phenomenon which has been named *Moore's law*, after Intel founder Gordon Moore who predicted this trend in the 1960s and 1970s. However, physical, heat dissipation and leakage current constraints are increasingly limiting our ability to keep packing ever more transistors into the same 2D planar area of silicon. While the industry is expanding this pursuit into the third dimension [81], it is also increasingly keen to find a fundamentally different approach to maintain – or even surpass – the exponential growth in computing capabilities.

Within quantum computing there remain many issues to resolve. On the physical side, it is extremely difficult to build a quantum computer that is able to maintain its computing elements reliably entangled over any significant length of time. This issue is known as maintaining *coherence* among the quantum elements, in comparison to a deterioration into *decoherence*, most often caused by stray electrical and magnetic noise factors. Current designs struggle to achieve longer coherence, even with the aid of cooling the system ambiance to near absolute zero, which means that managing such a system is physically demanding and expensive.

On the conceptual side of quantum computing, many questions remain open as well. A quantum computer cannot be programmed using the same methods as a classical computer. Designing an algorithm for a quantum computer boils down to composing just the right choreography[5] of quantum interferences among the computing elements to achieve the desired outcome, which is a fundamentally different way of programming, closer in kind to *analog computing*, which predates digital computing.

---

[4]Normally the improvement is sub-linear due to overheads, although in certain situations some amount of super-linearity can be achieved, depending on the overall system architecture and the application at hand.

[5]This elegant description is taken from an instructive conversation between Scott Aaronson and Lex Fridman [82].

5.5   How will AI and quantum computing shake hands?

Since both AI and quantum computing are still rapidly evolving, it is difficult to predict how the two will eventually interact; however, we can offer some speculation on the topic. One question of interest would be whether quantum computing can be useful for the computationally intensive training phase of deep learning. Training a large neural network requires vast amounts of data, and involves a large number of parameters, while quantum computing excels at complex calculations which can be compactly specified and parameterized, and the answer compactly read out. Every additional parameter requires additional qubits, and every data input decreases the computation's isolation needed for maintaining entanglement/coherence. Accordingly, we can venture the hypothesis that, unless a large dataset can be very economically encoded in a quantum computer, the latter will be of limited use for training large models, going forward. There is also the significant issue that quantum computers have a very limited output, so an application using a quantum computer for AI inferences would need to retrieve answers in a highly compact form. The size of the answer essentially needs to be at most as big as the number of qubits in the system [83]. Nevertheless, using quantum computers for machine learning ('Quantum ML') is an active field of study, pursued among others by CERN in the context of its quantum technology initiative[6].

A second question of interest is whether AI will enable the future of quantum computing. For one thing, it may help us write programs for quantum computers. As discussed above, designing quantum algorithms remains an art mastered by very few so far. Potentially, an AI model can be trained to translate certain categories of classical programs to the quantum domain. After all, translation problems of all stripes have revealed themselves to be a strong suit of deep learning in particular, ranging over natural language text, source code, audio and other data, so qubits and quantum gates may also enter its vocabulary. On the hardware side of things, the ability of AI to efficiently simulate quantum systems [84], [85] could provide a helpful resource to advance research on how to build a good quantum computer.

---

[6]https://quantum.cern/

There is also an aspect in which AI and quantum computing are in competition with each other. Both technologies require very substantial investments for research and development, and funds going to one are not available to the other. It has been argued that AI's fast progress is detrimental to investment in quantum computing, because AI is able to massively speed up and enhance classical computing, pushing out the time horizon for quantum computing to achieve 'supremacy', or at least economic feasibility, in any specific area [86].

Since AI and quantum computers have different strengths, it is likely that they will complement each other in the future, by acting as components in a larger system which can draw on both, according to the nature of the task at hand, in a hybrid computer architecture.

6

Why believe AI? The role of machine learning in science

Artificial intelligence has made significant inroads into science over the past decade, especially under the machine learning variant called *deep learning*. This chapter will examine the challenges and questions concerning the role of AI in the scientific process. We proceed by building on two articles by Naomi Oreskes, 'Why Believe a Computer?' [87] and 'The Role of Quantitative Models in Science' [88], written at the turn of the millennium. These articles critically examine the role that computers should play in science. Computer simulation has since become widely accepted as the 'third pillar of science' [89], alongside theory and physical experimentation. While AI will not replace any of these paradigms, it may enrich them, and perhaps be regarded as a fourth paradigm itself [90]. With the benefit of two decades' worth of hindsight, we revisit some of Oreskes' arguments and refurbish them for the era of AI.

6.1 Testability and complexity

As a historian of science, Oreskes [87] describes how scientific epistemology has evolved across the centuries, with the aim of identifying the proper place (if any) that should be granted to computers within it. A key notion arising early on is that *testing* lies at the heart of science. Since at least the 17th century, upon the case made by Sir Francis Bacon, many scientists agree that a theory should be testable, and tested to be proven correct. This notion was refined by Karl Popper in the 1930s. Popper held that a theory should indeed be testable (or *falsifiable*), yet could never be fully proven correct. It could only be shown to be in accord with the available experimental data, but not proven to remain so with respect to any future evidence. Pierre Duhem, in his 1906 publication 'The Aim

DOI: 10.1201/9781032710525-6   44

and Structure of Physical Theory', did not reject the need for testability, but argued that any theory could be modified and extended post hoc so as to accommodate new findings, and the more complex the theory, the easier it would be to extend. This argument adds justification for the *Occam's razor* principle widely embraced by science, according to which a simple theory is preferable to a complex one if their explanatory powers are similar; a simple theory is easier to test, and it is more obvious when one attempts to modify it so as to accommodate incompatible evidence.

The dilemma[1] that arises in the context of computer models is the following: the more faithfully a computer model represents a complex real-world system, the harder it is to test. Unfortunately, a complete description of a natural process, if even possible, would be extremely verbose and intricate. For instance, when expressed in a programming language, the number of lines of computer code in a global climate model can easily reach into the millions, which is at least one order of magnitude longer than the complete works of William Shakespeare (albeit much less pleasant to read), and it still only remains a rough approximation of the underlying processes.

On one level of abstraction, deep learning is a much simpler program than any Global Circulation Model. In fact, only a handful of basic mathematical functions are required for implementing even the largest deep learning models. However, deep learning contains an enormous number of parameters (millions or even trillions in latest trends), which all have an influence on the model's operation, so that the complexity is shifted from the computer code to the parameter space. This abundance of parameters, or degrees of freedom, seems to fly in the face of Occam's razor, and is one of the main points of contention concerning the admissibility of deep learning in the scientific toolbox. A quote attributed to John von Neumann reads, 'with four parameters I can fit an elephant, with five I can make him wiggle his trunk', and this sentiment permeates much of the scientific and statistical thinking which scientists and engineers will have been imbued with throughout most of their education.

The danger which is being alluded to is that of severe *overfitting* (cf. Section 2.5), that is, of having a model which is so plastic that it will contort itself to accommodate any input data, without capturing any of its essential characteristics, and consequently exhibiting poor generalization capabilities: its accuracy given previously unseen input will be poor.

---

[1]Oreskes uses the term 'paradox'

Traditionally, the advised approach for avoiding overfitting has thus steered statistical modeling strongly in the direction of using as few parameters as possible. Since deep learning opts to *maximize* rather than minimize the number of parameters in order to attain state-of-the-art prediction accuracy in a continuously growing list of domains, how has overfitting been resolved?

Information-theoretic studies of deep learning, going under the heading of 'information bottleneck' [91] suggest that the optimization procedure used in deep learning, called stochastic gradient descent (cf. Section 2.4), may be somewhat naturally immune to overfitting under certain assumptions about the statistics of the data being used. All the same, additional algorithmic techniques with the explicit aim of avoiding overfitting were also introduced. One such technique, called *dropout* [92] (cf. Section 2.9), developed in the lab of Turing laureate Geoffrey Hinton, randomly perturbs the model, and forces it to construct internal redundancy for robustness. This technique turned out to be highly effective in avoiding overfitting, and has been widely applied ever since the deep learning model 'AlexNet' described in Chapter 5 won the ImageNet Large Scale Visual Recognition Challenge in September 2012. This image classification competition had up to that point been dominated by bespoke hand-crafted models, and the surprise takeover arguably ignited the ongoing deep learning revolution. From this perspective, Hinton's dropout can be seen as an alternative principle to Occam's razor. But is it compatible with science?

6.2   The purpose of science

This leads us to step back and ascertain which goals we are pursuing in a given scientific endeavor. Oreskes [88] states that

> the purpose of modeling in science must be congruent with the purpose of science itself: to gain understanding of the natural world.

Achieving understanding may be the purest driver of science, but there are others. Obtaining reliable predictions without true understanding would certainly rank as a more desirable scientific outcome than the absence of both. Arguably, quantum mechanics is an example of a theory which provides extraordinary capabilities for prediction, without necessarily delivering a deep understanding of the natural world. Deep learning, at the time of writing, is certainly much better at delivering predictions than understanding.

Besides prediction, informing policy is another necessary goal of science, that is, providing sensible governance recommendations based on the current body of knowledge, weighed against estimated risks and benefits. In this area, deep learning may be less suitable for direct application, since assessing risk and benefit typically requires context and common sense, two dimensions in which AI still needs vast improvement. It may, however, be usefully applied to produce predictions which inform such recommendations.

Although other objectives exist, *understanding* definitely reigns supreme among desired outcomes of the scientific process - even flawed understanding may serve as a stepping stone. Understanding allows us to build upon, to theorize further implications, thereby motivating new experiments which may in turn reveal flaws in our (previously assumed) understanding, driving progress as a result. In this area, deep learning has not made many contributions[2]. Research into extracting understanding from deep learning models does exist, going under such headings as explainable AI (XAI) and interpretable AI, but these efforts are still in their early stages.

Causal models (Section 2.17) are an alternative approach to AI and may be more fundamental to science. They are championed, among others, by Judea Pearl, also a Turing laureate, who holds that 'all the impressive achievements of deep learning amount to just curve fitting' [94]. In other words, deep learning is focused on finding statistical relationships in high-dimensional data, without considering the causal relationships which give rise to them. Human understanding is closely tied to causal interpretations of the natural world, and hence our AI models ought to speak that same causal language in order for us to learn something from them. *Causal*

---

[2]In some cases, deep learning has revealed new concepts. For example, the Go playing model AlphaGo stunned experts by defeating the best human players, using strategies which defied established wisdom (built up by the extensive scholarship of the game of Go over centuries), thus providing new understanding and insight [93]. This could however be a lucky accident, in the sense that AlphaGo did not explicitly describe this strategy to humans, but merely used it in its play, and the strategy happened to be highly visible precisely because it violated simple principles that Go players took as gospel.

inference and *causal discovery* are highly active areas of research today, although not nearly as active as deep learning. Building a causal model for a given problem domain requires much more domain knowledge than does deep learning, and the software packages for causal models are mostly in research stages, whereas the deep learning software stack is now industrial-strength. Additionally, the causal approach lags far behind deep learning in the following fundamental aspect: it requires the modeler to define *what the variables are* (and for best results, specify any partial domain knowledge on the causal relationships between those variables). In contrast, deep learning's most significant productivity enhancer has been to figure out all by itself what the variables are, from raw data, eschewing the need for such 'feature engineering'. Unfortunately, the variables that deep learning comes up with don't necessarily have a causal interpretation. Perhaps, combining the best of both approaches will supply an answer; however, this will still require major conceptual breakthroughs.

6.3   High-dimensional output, low-dimensional internals

Returning to the concern of overfitting, the picture we may often implicitly have in mind is that of a very high-dimensional input, combined with a low-dimensional output. For instance, petabytes of physical measurements across time and space as a source for tens or hundreds of atmospheric, oceanic and other variables as input, and as output, the global average surface temperature. In such a setting, overfitting is indeed an overwhelming concern. The situation is qualitatively different, however, if the complexity of the output matches that of the input, for instance if a model were to get the same petabytes of input as described above, but now had to predict values for all those same variables for subsequent timesteps and myriad physical locations, as its output. Predicting a billion values correctly, based on a fundamentally flawed model, would be much more of a fluke than predicting just a single value or a single time series.

For such balanced high-dimensional ambitions between input and output to become practical, the availability of the required data can of course be a limiting factor; however, Oreskes' comment [87] that

> the availability of data from the natural world has not kept pace
> with advances in theory and computation

is being potentially turned on its head. The data volumes from Earth
Observation programs do not yet match those generated by model simulations, but the challenge is shifting from data dearth to data deluge,
as in many other scientific arenas as well. For example, the Copernicus
data archive, hosting the satellite observations collected by the European
Space Agency's Sentinel missions, is expected to grow from 34 to 80
petabytes within six years [95]. NASA's Earth Science Data Systems
(ESDS) is expected to grow to approximately 250 petabytes by the year
2025 [96]. To get a sense of scale, imagine 250 petabytes of written text,
printed out on A4 paper sheets. The stack of paper would reach far past
the moon's orbit, and cover nearly 10% of the distance from the Earth
to the sun.

A deep learning design pattern has emerged as a powerful way to improve
both the trainability and the interpretability of models: combining high-
dimensional input and high-dimensional output with a lower-dimensional
part in the middle. Some work along this vein has shown that manipulating parameter values within this lower-dimensional mid-section can
have meaningful or intuitive effects on the output[3], while manipulating
values in other layers typically results in more haphazard changes, which
indicates that the network is somehow distilling the task down to its
essence within the lowest-dimensional layers. Whether this representation
is in a language that we can truly understand remains an open question.

6.4 Data impedance mismatch and end-to-end DL

On the topic of data, Oreskes raises an important limitation of working
with computer models, which is that the data must typically first be
transformed in order to match the variables that were designed as inputs
to the models. This can be referred to as the data *ingest* problem. The
same applies on the output side of the process [87]:

---

[3]Some deep generative models such as variational autoencoders (VAEs) build on
this observation.

> The gap that exists between empirical input and model parameters
> is mirrored by a gap between model output and the data that
> could potentially confirm it.

In engineering terms, we could describe this as an impedance mismatch
between the data and the typical computer model. As it turns out, deep
learning has an ace up its sleeve in this respect, because it is able to
handle essentially any input and output we choose. The trend in many
areas where deep learning is applied to large datasets is to use an *end-to-
end* approach, from raw data directly to desired output. For instance, the
classical approach to the problem of speech recognition is to decompose
it into a sequence of sub-problems; first extract specific features from
audio data, then process those features to obtain phonemes/syllables,
then use these to construct words, and finally, assemble a full textual
transcript based on these words [97]. In contrast, the end-to-end deep
learning approach skips all these explicit intermediate representations,
and directly learns how to map raw audio to a transcript. One could
argue that this contributes to the black box aspect of deep learning.
However, since the use of a traditional model requires the production of
mappings from data to intelligible variables and from model output to
verification data, it should be noted that these same mappings would
also be available for the purposes of evaluating deep learning models.
Indeed, end-to-end thinking is being considered in scientific applications,
such as numerical weather prediction [98]. The related topic of generative
models is discussed in the upcoming Chapter 7.

Given the current uncertainties surrounding the future (and even present)
capabilities of deep learning, it is difficult to assign it a clear cut role
in the scientific process. On the one hand, the predictive capabilities
it provides are so useful that we may be compelled to rethink basic
principles in scientific thinking, such as Occam's razor. On the other, it
may be that efforts in understanding and interpreting the internals of
deep learning models will produce the type of understanding that we
classically expect from scientific models, allowing us to have our cake
and eat it too. Later in Chapter 8, we will cover a different kind of AI
that is fully aligned with the current paradigms of the scientific process:
causal models.

7

Generative AI

The deep learning revolution was largely sparked by successes in image classification, epitomized by the victory of the convolutional neural network AlexNet in the 2012 ImageNet competition. Decades of research in handcrafted feature detectors were disrupted by this model, trained on two consumer-grade GPUs, which radically changed the minds of computer scientists concerning neural networks. Prior to this incontrovertible demonstration, the consensus view was that neural networks were a subpar type of machine learning model. As far back as the 1960s, early mathematical results had shown that single-layer neural networks had severe limitations in the variety of functions that they could represent, which likely discouraged further inquiries into their potential. In 1989, it was mathematically proven that a neural network with just three layers (including the input and output layers) and a non-linear activation function is in fact a universal function approximator, meaning that given enough neurons, the model can approximate (nearly) any function arbitrarily well [99], [100]. Despite the ensuing revival in scientific interest, neural networks did not reach mainstream status at that time because they were difficult and expensive to train compared to other models. As discussed earlier in the book, the combination of abundant data and cheap computation were key in finally propelling neural networks to the fore.

Two limitations that beset machine learning up to that prior point have undergone a significant reexamination as a result. Firstly, previous ML models were task-specific; a model trained on a given task was essentially useless when applied to a different, even highly related task. Secondly, the output of an ML model was usually restricted to be low-dimensional, compared to the input data. Deep learning started out in much the same vein. The input to the AlexNet model is a 256×256 pixel color image, and each pixel can take on 2^{24} possible colors. This means that the number of possible inputs is over 10^{473479}, an astounding number given that the number of atoms in the universe is commonly estimated to be about 10^{80}.

DOI: 10.1201/9781032710525-7

The model's output, on the other hand, simply consists in one among 1000 object labels (such as 'Persian cat' or 'volleyball'). This is a very low-dimensional output as compared to the dimensionality of the input. It is also highly specific to the particular dataset of images that was compiled for the purposes of the ImageNet competition. In subsequent years, deep learning research has shown that low-dimensional output and task specificity were in fact not fundamental limitations, as we outline below.

7.1   Transfer learning and fine-tuning

It came as a surprise that image classification models trained on one task could in fact be repurposed for another, similar task, with modest computational effort. For example, a model trained on the ILSVRC 2012 dataset can be adapted to distinguish between an image of an American football and an image of a papaya, neither of which is a category in the original training dataset. It turns out that the weights learned in the early layers of the model are quite general-purpose feature detectors, which can be used as building blocks to detect nearly any kind of object. Only the last layers need to be retrained to adjust to a new object category. This technique is known as *transfer learning*, and requires much less computation and data than training the original model did. In the case of AlexNet, millions of images were required to train the model, and each image was 'looked at' 90 times over the course of the training (this is referred to as 90 *epochs*). By contrast, adapting the model to two new categories may require a few hundred additional images and a few epochs. Once the last layers have been retrained, while keeping the early layers frozen, one can also unfreeze the whole network and train a bit longer with a low learning rate, in order to further improve the final accuracy. This process is known as *fine-tuning*.

The success of early deep learning was thus achieved in supervised learning (cf. Section 2.2), where each data point comes with a label - the image category in the case of ImageNet. A drawback of supervised data is that it is labor-intensive to compile. For each datapoint, a label needs to be supplied, either by human annotation, or by some automatic process which itself needs human verification. The ImageNet images were labeled by enlisting tens of thousands of paid workers via crowdsourcing platforms [101], [102]. Because unlabeled data (in other words, simply data) is much more prevalent in the world than labeled data is, the ability to learn useful patterns from unlabeled data would obviously constitute an enormous advantage.

7.2 Unsupervised learning and generative models

Learning from unlabeled data is referred to as unsupervised learning, and it is the principal source of the power of generative deep learning models. Where the typical supervised ML model aims to learn a good mapping from X to Y, for example from images to category labels, an unsupervised model instead learns the joint probability distribution of X and Y, or simply the probability distribution of X if there are no labels Y. If the model is generative, it has the ability to sample from that learned distribution. In other words, it can *generate* new data points. If the data is a collection of cat pictures, the trained model will be able to produce a 'new' cat picture that was not part of the training data. Such unsupervised generative AI models have hugely improved in the past few years, and first gained prominence in the medium of text. The most emblematic generative models at the time of writing are the GPT family of large language models (LLMs) [103], [104], which are able to generate highly plausible text on nearly any given subject. GPT is the acronym for 'generative pre-trained transformer'. The 'transformer' part of the name refers to the transformer neural network architecture (cf. Section 2.12), and 'pre-trained' signifies that the model has been trained in an unsupervised fashion, and hence on a large body of text. Finally, the 'generative' prefix indicates that the model is able to produce new data from the distribution that it has learned, i.e. to generate new text. With additional supervised fine-tuning of the pre-trained model, using for instance reinforcement learning from human feedback, the model can be adjusted so as to generate conversational responses to the user's inputs, while attempting to avoid offensive, harmful or otherwise inappropriate output. The most famous such interface is known as ChatGPT, released by OpenAI in late 2022, but other offerings of similar nature have rapidly joined the market since then. In particular, several competitive open source models have been released, such as the Llama models [105], [106] by Meta, Gemma [107] by Google DeepMind, and BLOOM [108] through an initiative led by company HuggingFace. These models are being applied to the generation of a wide variety of text, from poetry to book summaries to programming source code. Some current limitations of such models will be briefly discussed below.

Text is not the only medium in which generative deep learning has made leaps of progress. Image generation has also come a very long way, especially in the 'text-to-image' variety; the user supplies a short textual prompt describing the image to generate, and the model creates one or several images derived from that prompt. In this medium as well, OpenAI has supplied the most popular interface, based on its DALL-E family of models [109]–[111]. Some notable image generation models on the open source front include Stable Diffusion and DeepFloyd IF [22] by company Stability AI. Note that many of the new large models are *multimodal*, meaning that they are able to deal with data from different media at once. For example, GPT-4 can understand both text and images as input, even though its generative capability is limited to text only. The new frontier of generative models is video generation. Lumiere [112] is a model released by Google in early 2024, which uses diffusion (see Section 2.15) and a U-Net architecture which downsamples video in both space and time, in order to generate new short videos based on input text and/or an input still image. A few weeks later, OpenAI announced its own video generation model called Sora [113], showing samples of minute-long footage of a much higher temporal consistency and realism than in the prior state of the art. Based on the pace at which generative models are advancing, it is very likely that there will be several more highly capable models released by the time this book is published.

7.3   Limitations

Generative models have a mass market appeal, and some of the models discussed in this chapter have already been accessed by over a hundred million users [114]. The curious public has been experimenting with this new technology, attempting to figure out what it can do for them, and to understand its strengths and limitations. We briefly go into several of these limitations in this section.

7.3.1 Hallucination

ChatGPT (and other LLMs) are most commonly accessed via a text interface, enabling a written chat conversation with the AI. As such, users initially expect it to behave similarly as their usual human conversation partners. However, an LLM differs in at least several important aspects. When speaking with a friend or colleague, we normally expect them to be truthful. In addition to this, we assume that they will adjust their level of confidence according to their knowledge of the subject being discussed, indicating by verbal and non-verbal cues when they feel competent to opine, and when they are less sure of themselves. Finally, we can generally expect them to be able to explain why they hold a certain belief, in particular by citing their sources of key pieces of information relevant to the topic. As it turns out, LLMs presently fall short on these expectations. An LLM will regularly generate textual statements that are completely false. In fact, it will do this in a highly confident tone, producing a mix of truthful and false statements with the aplomb of a con man. This has been referred to as *hallucination* and sometimes *confabulation.* When users realize that the model has invented something in the course of their conversation, they can naturally take a dislike to the technology, associating its response with a fabrication designed to mislead.

Indeed, LLMs are trained on enormous text datasets, containing a significant portion of the web, including discussion forums of all kinds. Surely, not all utterances on the web are truthful, and of course, much of the text that exists in the world contradicts itself in part, reflecting the huge diversity in human opinions and beliefs, along with their falsehoods and biases. Consider also that some text may have been accurate when it was written, but has become out of date, or is otherwise learned without the necessary context. For example, novels and fiction in general account for a sizeable chunk of published text, which should be distinguished from newspapers, or scientific literature, or online discussion forums. Therefore, even if the model was able to reproduce 'the web' with high fidelity, the user could get very different outputs, depending on which part of the web the model was tapping into. On top of this, an LLM only learns an approximation of this enormous amount of text, which adds another level of difficulty to the goal of keeping it truthful, and which would remain even if one had fully vetted the entire body of text used for training. Unfortunately, in its current design, an LLM is unable to trace back from its choice of words in generated text, to the portions of the training data that would have had the largest influence on that choice. Consequently, we can typically not determine why the model responds in a certain way to a given prompt, without clever detective work. If we ask it to explain itself, it will typically just hallucinate a response that sounds plausible, or apologize (as a somewhat generic response added by the service provider through fine-tuning or other means) and state that the mistake may be due to incomplete data. Finally, one should be aware that the process of generating text includes randomness. At every stage of inference, the model samples a random word (or token) based on a conditional probability distribution given the preceding words. This means that the text provided as an answer to any prompt is partly driven by rolls of numeric dice.

Where images are concerned, it appears that we have very different expectations of a generative model than we do in the case of text. We are aware that the model generates fictional images, which may contain inaccuracies and counterfactual features. Even though it may be able to depict familiar objects and famous persons, we are not surprised by hallucination in this context. In fact, here it is arguably considered as a feature rather than as a bug. Image generation services are seeing widespread adoption for creative purposes, and inaccuracies are often humorous or otherwise interesting. Of course, this also brings new potential for misuse, such as in the case of *deepfakes*, which consist in applying a generative model to produce an image or video using someone's likeness. Political commentators fear that deepfakes can be used to influence elections [115], [116], and indeed if the technology reaches such a level of quality that even forensic analysis is unable to differentiate between real and fake images, we will need to rethink the very standing of photographic, video and audio evidence in our epistemic processes. The ability to choose between factual and invented content is certain to be an important requirement in the future. Especially in a scientific application, models will need to contain mechanisms to guarantee that important aspects of the generated data are governed by applicable conservation laws or other physical constraints.

7.3.2   Bias

As was alluded to in the section on hallucination, a generative model learns to approximate the probability distribution of the dataset that it is trained on. If the dataset contains inaccuracies, falsehoods, and biases, the model will inherit these aspects of the data, unless specific care is taken to avoid them. For example, AI bias has been well documented in the domain of face recognition. Identifying a person based on a photograph of their face is a challenge that many research efforts have tackled. Since the 1990s, models have consistently performed best on the combinations of race and gender that were most represented in the training dataset [117], sometimes resulting in extremely poor accuracy when attempting to recognize a person with a racial minority background. The EU AI act, which the European Union's parliament and council have adopted in March of 2024, explicitly bans 'the use of real-time remote biometric identification systems in publicly accessible spaces for the purpose of law enforcement', unless some stringent legal criteria are met [118]. The main reason for this prohibition is based on considerations of fundamental privacy rights of citizens. If innocent persons got tangled up in police investigations simply because of a technological bias with respect to their immutable characteristics, it would be a severe injustice. The problem of bias[1] permeates all types of ML models, including generative ones. If a group is underrepresented in the training dataset, it is likely to be equally underrepresented in the generated data, unless the training procedure (or the downstream generative procedure) is adjusted to rebalance the representation. On the topic of generative models, the EU AI act stipulates that the use of generative models for producing deepfake image, audio or video content, must be disclosed. Likewise, the use of generative models for textual output must be disclosed if the text is published with the purpose of informing the public on matters of public interest, unless a natural or legal person reviews the material and holds editorial responsibility for the published content [119].

7.3.3 Retaining copies and collapsing diversity

Several kinds of text-to-image generative models have been shown to retain a nearly exact memory of a small subset of the training data. In consequence, upon inputting the right text prompt, the model 'generates' an image that is nearly identical to one of the images in the training dataset [120]. In most cases, this is an undesirable aspect of the model, causing issues in terms of privacy, intellectual property rights, or lack of creative ability. A similar phenomenon has been observed in the case of LLMs. In particular, when asked to repeat a given word forever (as a

---

[1]To clarify: most neurons in a neural network have a parameter called the bias term (cf. Section 2.7), which is unrelated to the notion of bias under discussion here.

prompt), ChatGPT repeated that word many times, and then all of a sudden, began outputting some of the text it had been pre-trained on, verbatim [121]. These discoveries provide a clear indication that the way in which deep generative models learn is not well understood yet.

Current generative models also seem to be unable to generate as much diversity as is present in their training datasets. This was noticed when using a dataset of images generated by a model to train a second model, then in turn using the output of the latter to train a third, etc. Each successive dataset obtained in this way tends to be less diverse than the previous one. This has been referred to as 'model collapse' [122], and has been observed in text [123] and images [124]. From a statistical point of view, there is evidence that 'rare' features are discarded over successive generations, thereby increasingly converging on average or majority features. This would be akin to the model creating its own, ever more narrow view of the world.

7.4   Implications for the Earth sciences

So far, generative models have not been broadly used in scientific applications, with a notable exception in numerical weather prediction (NWP). Here, three deep learning weather models were published in close succession, within the span of one year. Nvidia built FourCastNet, a neural network based on Fourier neural operators, doing much of its work in the frequency space instead of in pixel space [46], [125]. Google released its graph neural network model called GraphCast, which operates on a graph composed of the vertices of a globe-spanning mesh [47]. Finally, Huawei announced its Pangu-Weather model, built on a 3D Earth-specific transformer neural architecture [48]. All three models used the ERA5 global atmospheric reanalysis dataset in their training procedure, and were made to autoregressively predict the new state of global weather from the previous state. They were compared to the best-in-class classical numerical weather prediction models, which operate by numerically solving partial differential equations describing state transitions. Each neural network model outperformed the classical ones in several important metrics, disproving a recent assessment that a number of fundamental breakthroughs were likely to be required before achieving this feat [98]. In particular, Pangu-Weather showed higher accuracy across the board, in predicting geopotential, temperature, specific humidity, and wind speed, at the surface height level of 500 hPa. Nevertheless, a classical physics-based model was employed for the reanalysis that allowed the production of the training data, and therefore, classical models remain a critical requirement for the success of AI/ML in NWP, as well as in many other applications. It should also be noted that the three AI/ML models did not perform well in predicting some important aspects of the 2023 storm Ciarán, such as the maximum wind speeds at 10 meter height [126]. Generally speaking, AI/ML models will perform much better at interpolation than at extrapolation. When faced with a situation that lies far outside of their training data, they may predict with much lower accuracy than a physics-based model would.

Generative models are also being pursued in remote sensing. Instead of manipulating discrete language tokens, some research streams attempt to use same technology on multispectral and multi-instrument satellite observations. In this context, a token is a pixel with several layers, each layer stemming from a specific band in a satellite image. The abundance of such pixel data in the remote sensing archive is such that autoregressive self-supervised approaches akin to those used in large language models are feasible, and have led to promising predictive performance for several downstream tasks, such as predicting future surface reflectances across the 400-2300 nanometer range [127]. Another generative model architecture, called generative adversarial[2] network (GAN), has previously been applied to Earth observation tasks. For example, conditional GANs (CGANs) were used to fill in voids in incomplete satellite observations, such as mountain shadows in incomplete radar data, as well as for spatial interpolation and image pansharpening [128].

7.5 Outlook

The potential of generative models is still being explored with much excitement, especially given their success in the language modeling space.

---

[2]The word *adversarial* alludes to the fact that a GAN is trained as a pair of adversaries playing a game against each other: the generator and the discriminator. The generator produces a sample (e.g. an image), and the discriminator needs to distinguish between generated samples and actual datapoints. Both of these parts are trained in tandem, helping each other improve.

Some voices argue that they are a red herring. Yann LeCun, the deep learning pioneer who shared the 2018 Turing award with Geoffrey Hinton and Joshua Bengio, believes that generative models as they are currently designed, especially for image sequences, do not have the right approach to succeed in the long run. The strategy thus far has been to degrade the input by adding noise or occluding parts of it, and then learning to reconstruct all of the pixels. However, in many situations this is not a good objective, because there is not enough information to reconstruct the entire image exactly. According to LeCun, instead of focusing on the input distribution, one should aim to achieve high-quality reconstruction in an abstract space [129]. The JEPA proposal from his group is to produce a joint embedding of an image and of a noisy version of that image, and to learn to reconstruct the latent representation of the original image from the latent representation of the noisy one [130].

It is very much an open question as to what extent neural network based systems will be able to perform complex reasoning. DeepMind's AlphaGeometry model [131] is reported to have achieved top-ranking performance in geometry problems, solving 25 out of 30 Olympiad-level geometry test problems, which is nearly as good as the average of gold medalists of the International Mathematical Olympiad. It is a neuro-symbolic system, using a neural language model, and a symbolic deduction engine, working in concert. Essentially, the neural network generates possible geometric constructions, and the symbolic deduction engine attempts to use them to obtain the desired outcome. The model produces human-readable proofs, that is, it is able to 'show its work'. Nevertheless, the deduction part is still done in a rather 'brute force' fashion, unlike how a human would go about the task, using intuition to guide the flow of the proof as well. Causal thinking is very much part of our intuitive thinking, and thus the ability to consider causal relationships while reasoning is likely to be an important milestone on the way to more advanced levels of intelligence. A workshop of causal researchers has found evidence that large language models were not consistently able to perform causal reasoning, and suggested the term 'causal parrots' [132] to describe them, building on an earlier critique that LLMs can behave like 'stochastic parrots', haphazardly stitching together linguistic forms they've observed in their training data [133]. In the next chapter, we provide an introduction to causal models: a branch of AI that aims to stay grounded in causal reasoning.

8

Causal models: AI that asks 'why' and 'what if'

As deep learning scales up at a steep exponential rate, in both industry and academia, another branch of AI is currently much less prevalent, yet it may be just as fruitful in the long run, if not more so. The paradigm goes under the umbrella term of causal models, and has been championed among others by Judea Pearl, who won the 2011 ACM Turing award for 'fundamental contributions to artificial intelligence through the development of a calculus for probabilistic and causal reasoning'. This chapter describes some fundamental concepts of causal models, and highlights some of its uses in studying Earth systems.

8.1 Causation vs correlation

The fundamental distinction between standard machine learning and causal modeling is that the former is based on correlation, while the latter focuses on causes and effects. Correlation refers to a statistical association between two variables, and students worldwide are taught that correlation does not imply causation. When we notice that two variables X and Y move in tandem, we know that this may be due to a number of reasons. It could be that X has a causal effect on Y. It could also be that Y has a causal effect on X. Likewise, it may be that a *common cause* W has a causal effect on both X and Y. In this case, W is often called a *confounder*. More subtly, it could be that the dataset under study has a *selection bias*, which can arise when its datapoints are selected according to a *common effect* Z of X and Y. Such a selection bias can distort the statistical relationship between X and Y. For example, if we wished to analyze the causal effect of sprint training on leg muscle mass, then using a dataset containing only Olympic sprinters would be a poor choice; that dataset has a strong selection bias, as it focuses on the

DOI: 10.1201/9781032710525-8

individuals with the world's very fastest running speeds, and running speed is causally influenced by both sprint training and leg muscle mass.

This is an example of what is referred to as a *spurious correlation* in the causality literature: a statistical association between variables that does not coincide with the causal relationship between these variables. Note that a correlation could also be accidental, which is popularly referred to as a spurious correlation as well, although its meaning differs from the one outlined above. An example of such a correlation is the one observed between the volume of Google searches for 'zombie' and the number of real estate agents in North Dakota, for the years 2004-2022, shown in Figure 8.1. This correlation would very likely disappear if we used a larger dataset. In the rest of this chapter, we will assume that sample sizes are large, in order to focus on the main concepts of causal models. However, it should be kept in mind that any limitations of classical statistical methods, when used on small samples, would also apply in the case of causal models.

Many researchers argue that standard machine learning models learn the statistical patterns present in the data, but do not have a concept of the causal underpinnings that gave rise to the data. Due to their black box design and large size, it is difficult to check whether deep learning models build internal representations of causal concepts or not. For example, large language models do manifest some level of apparent causal reasoning ability. However, they are still comparatively weak at such reasoning tasks, so it is plausible that the bit of causal reasoning they do exhibit, simply mimics similar or analogous text in the training data. Indeed, as was previously mentioned, there are reasons to suspect that they are just 'causal parrots' [132]. Causal models, on the other hand, are built on a completely different premise. They are explicitly designed to incorporate causal knowledge and assumptions, and to answer causal questions.

8.2   Causal graphs

Causal methods were developed within multiple different disciplines, from econometrics to epidemiology to computer science, and therefore they come in many variations. One practice that has found widespread

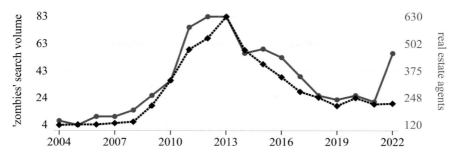

FIGURE 8.1 Example of a spurious correlation in the 'accidental' sense. The relative volume of Google searches for 'zombies' is highly correlated with the number of real estate agents in North Dakota (data sources: Google Trends, US Bureau of Labor Statistics), with an r value of 0.936. This correlation, and a sizeable collection of others like it, were obtained by *data dredging* a large database of time series: testing many possible pairwise combinations of time series, over many time segments and intervals. When one does this without adjusting the threshold for statistical significance (e.g. using the Bonferroni correction [134], [135]), many false positives can result. This is also referred to as *p-hacking* in some contexts. The collection of humorous correlations is curated by Tyler Vigen at http://www.tylervigen.com/spurious-correlations.

adoption is the use of *causal graphs*[1], also referred to as *causal diagrams*. A causal graph is a graph in which nodes represent variables, and edges represent causal influences. If there is an edge from node X to node Y, it signifies that X causally affects Y, in the sense that if we were to surgically change X, then Y would change in response. Pearl describes this by stating that 'Y listens to X' [24]. For example, consider the causal graph in Figure 8.2. It describes causal relationships between the insolation I a tree receives from the sun, the tree's growth G, and

---

[1]The material on causal graphs presented here mainly derives from the approach known as structural causal models (SCM). Although SCMs are not presented in this book, an SCM always implies a causal graph. For a comprehensive treatment of SCMs, the reader can consult references [25], [136]. Likewise, the main alternative approach to causality, called *potential outcomes* [137], is not covered in this book.

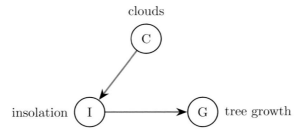

FIGURE 8.2 Example of a causal graph. According to this graph, insolation I has a causal effect on tree growth G, and clouds C have a causal effect on I.

the presence of clouds C. Specifically, it posits that insolation causally affects growth, and that clouds causally affect insolation.

The graph can be used as a means to communicate a hypothesis, which is readily understood by humans, and can serve to make questions and assumptions explicit. For instance, a colleague may point out that in order to model the situation more accurately, a direct edge from C to G is also needed, because sufficient water is necessary for tree growth, and clouds are necessary for rain. In other words, they may propose the causal graph in Figure 8.3 as an alternative hypothesis. Of course, one could add more variables, such as rain and soil moisture, in order to provide a causal description that is appropriate for the research question at hand, especially if data for these variables is also available.

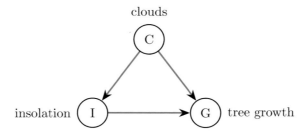

FIGURE 8.3 Alternative causal graph to Figure 8.2, in which a direct edge from clouds C to tree growth G is added, in order to model the effect of rain.

8.3 Causal inference

In addition to being human-readable, a causal graph is also machine-readable. It can be provided as a hypothesis to a causal inference engine, together with data. The engine is then able to check whether the graph is plausible with respect to the data, or whether the graph must be rejected. Furthermore, the engine can use the data and the graph to fit a quantitative model, which can then be used to answer queries about the system. In this instance, an example query could be, 'if we decrease insolation by 10%, what change will we see in tree growth?' An important point here is that this query can in general not be answered simply by looking at the conditional distribution of tree growth given insolation. There is a key difference between G given that we *observe* I, and G given that we *impose* I (i.e., that we intervene so as to force I to take on a certain value or distribution). For instance, suppose we intervened on insolation I in Figure 8.3, by blocking off the sunlight to the tree with a large screen or wall, and shining a large electric light on the tree, with an equivalent light spectrum, but using an intensity of our choosing. With this setup, we could now set the intensity so that it 'replays' the insolation observed during the previous week. The tree growth occurring with this chosen artificial insolation may be quite different from the tree growth that was observed for the previous week, because the causal effect of clouds on tree growth still remains, and we have not intervened on this variable C. Effectively, by intervening on insolation I, we transform the causal graph into what is shown in Figure 8.4, where the edge from clouds to insolation has been removed. Indeed, we've severed this causal influence, freely choosing the insolation.

The *do-calculus* is an important mathematical formalism for handling the distinction between observational and interventional probabilities. Its key innovation is the addition of the *do*-operator, which allows one to express an intervention. For example, $do(I = i)$ corresponds to the intervention of setting insolation I to the value i. The expected value of tree growth G under this intervention can be written as $E[G|do(I = i)]$, and as discussed, it is not the same as $E[G|I = i]$ in general. The do-calculus provides a complete set of rules for converting between causal quantities

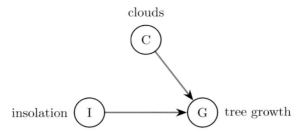

FIGURE 8.4 Modified causal graph from Figure 8.3, where we've intervened on insolation I. Since there is no longer a causal influence of clouds on insolation, the edge from C to I is removed.

and statistical quantities. For example, based on the graph in Figure 8.3, the statistical quantity obtained through do-calculus is the following[2]:

$$E[G|do(I = i)] = E_C[E[G|I = i, C]]. \qquad (8.1)$$

Remarkably, in this case, we are able to answer a causal query using only observational data. If we have good reason to believe that the causal graph is correct, then we need not perform an experiment requiring the expenses of a large wall and a gigantic floodlight. Of course, the result is only as valid as the causal assumptions.

8.4   Assumptions and limitations

Assumptions are a cornerstone of causal inference. Some of these assumptions are testable, others are not. For instance, assumptions in the form of a causal graph are usually testable, if appropriate observational or experimental/interventional datasets are provided. Other assumptions can be difficult or impossible to test. For example, an assumption that is frequently made in causal inference is that there are no hidden confounders, i.e. that the causal graph captures all variables that are causally relevant to the question being posed, also known as the causal

---

[2]This specific formula is called the *backdoor adjustment*, because it adjusts for the non-causal association leaving through the 'backdoor' of I, reaching G via C. The adjustment is obtained by conditioning on the confounder C.

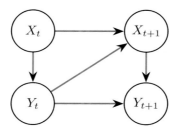

FIGURE 8.5 Causal graph containing a feedback loop between X and Y, even though the graph is acyclic. X_t represents the value of X at discrete time t, and equivalently for Y_t and Y. In this example, there is an instantaneous causal effect of X_t on Y_t, and a lagged causal effect from Y_t on X_{t+1}. Here each variable also causally influences its own future value.

sufficiency assumption. Another common assumption is that there is a one-to-one correspondence between the causal graph's structure and the conditional independencies that exist in the joint distribution over all its variables (technically called the Markov and faithfulness assumptions). The stronger the assumptions, the further the inference engine can leap, but the more justification needs to be supplied as to why the assumptions are reasonable in the context of what is being modeled. In this sense, there is a tradeoff between the strength of the assumptions, and the believability of the conclusions.

In many causal methods, one assumes that the causal graph is acyclic, i.e. contains no directed cycles, in the sense that walking the graph from node to node along the edges, in the direction of the arrows, never leads to revisiting a previously visited node. This seems like a strong limitation in the context of Earth systems, where feedback loops abound (cf. Section 4.4). However, methods for the causal analysis of time series can deal with this by using time indices for each relevant variable. See for example Figure 8.5, where a simple feedback loop between two variables X and Y is depicted. Even though the graph itself contains no cycles, it is able to represent a mutual causal dependence between X and Y across time.

8.5 Causal discovery

A part of causal research is aimed at 'causal discovery': ferreting out the causal graph from data. This can require a lot of data, and it is not always possible to obtain the full graph. For instance, causal discovery methods that are based on detecting conditional independence can only discover a set of graphs, called a Markov equivalence class, which in some cases could contain just a single graph, and in other cases contain multiple possible graphs that disagree on the direction of some of the edges. Many discovery methods are also unable to detect some subtle types of dependence. For example, if variables X, Y and Z are statistically dependent on each other, however they are pairwise independent (that is, X is independent of Y, X is independent of Z, and Y is independent of Z)[3], then most methods would prematurely and incorrectly conclude that there is no causal link between X, Y and Z. Of course, if we are able to obtain interventional data, e.g. by performing an experiment that intervenes on some of the system's variables, our possibilities of causal discovery are increased.

As discussed in the previous section, making additional assumptions allows for more powerful causal inference, as long as those assumptions are justified. This also applies to causal discovery, where assumptions about the functional form of dependendance between the variables, as well as about the statistical distributions of noise, are very helpful. The linear Gaussian setting is the worst case for causal discovery, which may come as a surprise, since it is one of the simplest cases from a statistical modeling perspective. However, non-linear causal relationships between variables are often easier to detect than linear ones, as the non-linearities allow more aspects of the data-generating mechanisms to be identified [138]. Similarly, in a linear setting, if we have good reason to assume that the noise distributions are non-Gaussian, then causal discovery can be achieved where it is otherwise not feasible, e.g. using a linear non-Gaussian acyclic model (LiNGAM) [139]. Causal discovery is seeing an increasing number of applications in the Earth sciences, for instance in analyzing climatological time series. In particular, applying the PCMCI causal discovery algorithm [140] to surface pressure anomalies in the West Pacific and to surface air temperature anomalies in the Central Pacific and East Pacific yields a causal graph that identifies the well understood Walker circulation. A similar technique used for the Arctic climate detects that Barents and Kara sea ice concentrations are important drivers of mid-latitude circulation,

---

[3]For example, consider the case when X and Y are independent coin flips (heads/tails), and Z is true if the total number of heads equals 1, and false otherwise. The variables are pairwise independent, but each variable is fully defined given the two others.

influencing the winter Arctic Oscillation through multiple causal pathways. Readers are referred to Runge et al [23] for a comparison of several causal discovery approaches for time series in Earth system science.

8.6 Interactions with machine learning and deep learning

For the time being, there is only limited interaction between causal research and the mainstream machine learning efforts focused on deep learning. For example, at Nvidia's GTC conference in March 2024, out of a total of 337 online sessions, 261 matched the search term 'AI', and 130 matched 'generative', whereas searching for 'causal', 'caused' and 'causality' yielded one, one and zero results, respectively. Although causal models and deep learning are often not straightforward to combine[4], achieving a synergy between the two would be highly valuable. They are somewhat akin to Kahneman's fast and slow modes of thought [142]. Deep learning resembles our fast and instinctive 'System 1', with its remarkable abilities, but also replete with faulty shortcuts and biases. Causal models are closer to the slow and methodical 'System 2', using logic and careful deliberation based on facts and hypotheses. The combination of 'System 1' and 'System 2' is inarguably very useful to us humans, and quite possibly, a fusion between deep learning and causal models would represent a significant advance in AI.

While the two approaches have not yet been assembled into a holistic framework, there are some efforts to use deep learning and other ML techniques for the purposes of causal inference, causal discovery and discovery of equations from data. For instance, a technique known as 'double machine learning' can be used to adjust for confounding [143]. In the context of the example in Figure 8.3, this would amount to using ML to learn to predict I from C, and separately to learn G from C. Finally, an additional ML model then learns to predict $(G - \hat{G})$ from $(I - \hat{I})$, where $\hat{G}$ and $\hat{I}$ are the predicted values of G and I, respectively, obtained in the previous step. Combining these pieces allows us to remove the confounding effect of C, and therefore to obtain an unbiased estimate of the causal effect of I on G.

---

[4]The meta-learners framework [141] provides a possible approach to combine deep learning and causal models, allowing the use of deep learning as base models.

On the research front of discovering invariants and equations from data, an approach called *symbolic regression* is used to search a space of mathematical expressions in order to find an equation or formula that describes the data accurately. Deep learning and reinforcement learning in various forms can make this search more efficient. For example, a recurrent neural network can be used to generate the expressions, in a framework called *deep symbolic regression* [144]. This approach can be further optimized by ensuring that only physically meaningful expressions are evaluated, as guided by the physical units of the variables [145]. The subfield of discovering equations from data has received significant attention of late, and the interested reader is referred to a recent review article [146].

9

Conclusion

As this book attempts to convey, artificial intelligence is evolving very quickly, and in many directions all at once. Despite its flaws and shortcomings, it is plain to see that AI provides previously unavailable capabilities. The judicious and innovative use of these capabilities has the potential to transform a wide range of workflows in business, in the arts, and in the sciences. Therefore, it is advisable to stay well informed on the evolution of AI in all its forms. The online version of this book, accessible at the URL https://book.aiml.earth, aims to provide readers with updates on important or promising developments in this space.

Future additions to the website will include coverage of the concept of artificial general intelligence (AGI), which describes a hypothetical AI that reaches a level of cognition comparable to that of humans, across a broad range of tasks. We will also cover the ongoing legal question as to whether the training of models on copyrighted material constitutes an infringement of that copyright, which will have far-reaching implications, in particular in the space of generative AI. Both the online version and the back matter of this book contain a collection of resources, pointing to programming tools/frameworks and datasets that facilitate getting started with machine learning, or exploring a new facet of AI.

DOI: 10.1201/9781032710525-9

Resources

Data is a key resource for creating successful machine learning applications. In particular, benchmark datasets are necessary, which is being addressed in the Earth sciences [147]. The availability of open source models and source code is also very helpful, to learn about previous efforts, and avoid reinventing the wheel. Finally, it is important to interact with a community of fellow practitioners, to find advice and potential partnerships.

Some helpful resources in these regards are listed below, in alphabetical order.

Climate Change AI

Web page: https://www.climatechange.ai/

Climate Change AI is an initiative that aims to catalyze impactful work, at the intersection of AI (especially machine learning) and climate change. In a recently refreshed version of the open access paper 'Tackling climate change with machine learning' [1], the authors collectively catalog and rank many concrete opportunities for ML to make a tangible difference in the quest for climate change mitigation and adaption, in electricity systems, transportation, buildings and cities, industry, farms and forests, carbon dioxide removal, and other domains.

ClimateSet

Web page: https://climateset.github.io

Climate models are important tools for analyzing climate change, and

73

predicting its future impacts. ClimateSet [148] assembles a collection of inputs and outputs from 36 climate models from Input4MIPs and CMIP6, and makes them available in a conveniently preprocessed form for large-scale ML applications. It includes inputs and outputs for five SSP scenarios, four forcing agents, and two climatic variables: temperature and precipitation.

Earth on AWS

Web page: https://aws.amazon.com/earth/

Amazon Web Services (AWS) is a very large cloud computing platform, which provides computational resources of many different kinds. The platform hosts copies of several important Open Data datasets, including data from the Sentinel and Landsat satellite missions. In addition, Amazon has an open Call for Proposals for research using the Earth on AWS datasets for building scientific applications, and successful proposal can be granted cloud usage credits.

Earth System Science Data

Web page: https://www.earth-system-science-data.net/

Earth System Science Data (ESSD) is an international journal for the publication of articles on original and high-quality, well-documented research data, with a clear emphasis on open access [149]. ESSD also supports a 'living data' process to support evolving datasets, which are subject to regular updates or extensions.

Google Earth Engine

Web page: https://earthengine.google.com/

Google Earth Engine is an online platform for geospatial analysis and computations, which is provided free of charge for academic and research use. It contains several petabytes of curated datasets, including satellite data from USGS/NASA and ESA, all accessible through a common API. The portal contains a development environment in which source code and interactive maps are displayed side-by-side.

Hugging Face

Web page: https://huggingface.co/

Hugging Face fosters an AI community, which builds, trains and deploys state of the art models, using open source tools in machine learning. The portal contains a repository of models, a collection of datasets, a set of demonstration apps, and a suite of collaboration tools.

Kaggle

Web page: https://www.kaggle.com/

Kaggle is an online platform for hosting machine learning competitions, which has built a large community of data scientists and ML practitioners. It is open to any type of data, and contains many Earth-related entries, such as the 'Understanding Clouds from Satellite Images' competition, by the Max Planck Institute for Meteorology, the 'LANL Earthquake EDA and Prediction' competition, by Los Alamos National Laboratory, and others. It is also a place where datasets are catalogued, so as to be found more easily by the community. For instance, it hosts the 'Climate Change: Earth Surface Temperature Data' from Berkeley Earth.

ML4Earth

Web page: https://ml4earth.de

Machine Learning for Earth observation (ML4Earth) is a German national center of excellence, led by the Technical University of Munich. Among its activities, ML4Earth maintains a collection of benchmark data products, which consist in pre-labeled EO datasets and baseline/pre-trained AI models. This enables a researcher get up and running more quickly, when tackling a new EO task. The EarthNets platform [150] maintained by ML4Earth contains a categorization of over 400 EO datasets.

Pangeo

Web page: https://pangeo.io

Pangeo is a community working collaboratively to develop software and infrastructure, in order to facilitate research in Big Data geoscience. The shared objective is to build an ecosystem of mutually compatible, open source geoscience software packages, following established best practices in the scientific python community.

Radiant MLHub

Web page: https://mlhub.earth/

The Radiant Earth Foundation is a non-profit foundation, whose goal is to increase the positive impact of Earth Observation through machine learning. The Radiant MLHub brings together training data, models, and a community of participants with backgrounds in EO, geospatial data, and machine learning.

Sentinel Hub

Web page: https://www.sentinel-hub.com/

Satellite data from many Earth Observation programs, such as the Sentinel missions (ESA/Copernicus) and Landsat missions (NASA/USGS), are available at Sentinel Hub under a unified API. The website also contains a graphical 'EO Browser' which allows a visual exploration of datasets that are available for a given time span and geographical extent.

SpaceML

Web page: https://spaceml.org/

SpaceML is a machine learning toolbox, and a developer community that builds open science AI apps, for space science and exploration. It is part of the Frontier Development Lab (FDL), supported by NASA, DOE, and ESA.

WeatherBench 2

Web page: https://sites.research.google/weatherbench

In order to facilitate global weather forecasting with ML, in particular for the medium-range timeframe (1-14 days), Google Research has published the WeatherBench framework (now version 2). The framework enables the evaluation and comparison of various weather forecasting models, using open-source evaluation code. It also contains ground-truth and baseline datasets [151].

Bibliography

[1] D. Rolnick *et al.*, "Tackling Climate Change with Machine Learning," *ACM Computing Surveys*, vol. 55, no. 2, pp. 42:1–42:96, Feb. 2022, doi: 10.1145/3485128.

[2] N. L. Hickmon, C. Varadharajan, F. M. Hoffman, S. Collis, and H. M. Wainwright, "Artificial Intelligence for Earth System Predictability (AI4ESP) Workshop Report," Argonne National Lab. (ANL), Argonne, IL (United States), ANL-22/54, Sep. 2022. doi: 10.2172/1888810.

[3] A. McGovern and A. J. Broccoli, "Editorial," *Artificial Intelligence for the Earth Systems*, vol. 1, no. 1, Jan. 2022, doi: 10.1175/AIES-D-22-0014.1.

[4] A. McGovern, "Creating trustworthy AI for Weather and Climate." Feb. 2024. Accessed: Mar. 28, 2024. [Online]. Available: https://www.youtube.com/watch?v=n99yWkrvx2s

[5] S. Russell and P. Norvig, *Artificial Intelligence: A Modern Approach*, 4th ed. Hoboken: Pearson, 2020.

[6] C. M. Bishop, *Pattern Recognition and Machine Learning*, 1st ed. New York: Springer, 2006.

[7] F. Fleuret, *The Little Book of Deep Learning*. lulu.com, 2023. Accessed: Feb. 01, 2024. [Online]. Available: https://fleuret.org/public/lbdl.pdf

[8] I. Goodfellow, Y. Bengio, and A. Courville, *Deep Learning*. MIT Press, Cambridge, MA 2016.

[9] G. Hinton, "Coursera Neural Networks for Machine Learning lecture 6," 2018. Available: https://www.cs.toronto.edu/~tijmen/csc321/slides/lecture_slides_lec6.pdf

[10] J. Duchi, E. Hazan, and Y. Singer, "Adaptive Subgradient Methods for Online Learning and Stochastic Optimization," *Journal of Machine Learning Research*, vol. 12, no. 61, pp. 2121–2159, 2011, Accessed: Feb. 09, 2024. [Online]. Available: http://jmlr.org/papers/v12/duchi11a.html

[11]   D. P. Kingma and J. Ba, "Adam: A Method for Stochastic Optimization," Jan. 29, 2017. http://arxiv.org/abs/1412.6980 (accessed Feb. 09, 2024).

[12]   D. Maclaurin, "Modeling, Inference and Optimization with Composable Differentiable Procedures," PhD thesis, Harvard University, Graduate School of Arts & Sciences, Cambridge, MA, USA, 2016. Available: http://nrs.harvard.edu/urn-3:HUL.InstRepos: 33493599

[13]   K. Fukushima, "Visual Feature Extraction by a Multilayered Network of Analog Threshold Elements," *IEEE Transactions on Systems Science and Cybernetics*, vol. 5, no. 4, pp. 322–333, Oct. 1969, doi: 10.1109/TSSC.1969.300225.

[14]   X. Glorot, A. Bordes, and Y. Bengio, "Deep Sparse Rectifier Neural Networks," in *Proceedings of the Fourteenth International Conference on Artificial Intelligence and Statistics*, Jun. 2011, pp. 315–323. Accessed: Feb. 10, 2024. [Online]. Available: https://proceedings.mlr.press/v15/glorot11a.html

[15]   Latent Space podcast, "Ep 18: Petaflops to the People with George Hotz of tinycorp." Jun. 20, 2023. Available: https://www.youtube.com/watch?v=K5iDUZPx60E

[16]   K. Fukushima, "Neocognitron: A Self-organizing Neural Network Model for a Mechanism of Pattern Recognition Unaffected by Shift in Position," *Biological Cybernetics*, vol. 36, no. 4, pp. 193–202, Apr. 1980, doi: 10.1007/BF00344251.

[17]   Y. LeCun and Y. Bengio, "Convolutional Networks for Images, Speech, and Time Series," in *The Handbook of Brain Theory and Neural Networks*, Cambridge, MA, USA: MIT Press, 1998, pp. 255–258.

[18]   A. Vaswani *et al.*, "Attention Is All You Need," arXiv, Dec. 05, 2017. doi: 10.48550/arXiv.1706.03762.

[19]   A. Dosovitskiy *et al.*, "An Image Is Worth 16x16 Words: Transformers for Image Recognition at Scale," Jun. 03, 2021. http://arxiv.org/abs/2010.11929 (accessed Feb. 20, 2024).

[20]   L. Ouyang *et al.*, "Training language models to follow instructions with human feedback," Mar. 04, 2022. http://arxiv.org/abs/2203.02155 (accessed Feb. 20, 2024).

[21]   J. Ho, A. Jain, and P. Abbeel, "Denoising Diffusion Probabilistic Models," Dec. 16, 2020. http://arxiv.org/abs/2006.11239 (accessed Feb. 20, 2024).

[22]   R. Rombach, A. Blattmann, D. Lorenz, P. Esser, and B. Ommer, "High-Resolution Image Synthesis with Latent Diffusion Models," Apr. 13, 2022. http://arxiv.org/abs/2112.10752 (accessed Jan. 30, 2024).

[23]   J. Runge *et al.*, "Inferring Causation from Time Series in Earth System Sciences," *Nature Communications*, vol. 10, no. 1, p. 2553, Jun. 2019, doi: 10.1038/s41467-019-10105-3.

[24]   J. Pearl and D. Mackenzie, *The Book of Why: The New Science of Cause and Effect*, 1st ed. USA: Basic Books, New York 2018.

[25]   J. Peters, D. Janzing, and B. Schölkopf, *Elements of Causal Inference*. MIT Press, Cambridge MA 2017.

[26]   M. Sonnewald, C. Wunsch, and P. Heimbach, "Unsupervised Learning Reveals Geography of Global Ocean Dynamical Regions," *Earth and Space Science*, vol. 6, no. 5, pp. 784–794, Mar. 2019, doi: 10.1029/2018EA000519.

[27]   K. Tsipis, "Machine Learning Identifies Links between World's Oceans," *MIT News | Massachusetts Institute of Technology*. https://news.mit.edu/2019/machine-learning-identifies-links-between-world-oceans-0320, Mar. 2019.

[28]   L. van der Maaten and G. Hinton, "Visualizing Data Using t-SNE," *Journal of Machine Learning Research*, vol. 9, no. 86, pp. 2579–2605, 2008, Accessed Feb. 01, 2024. [Online]. Available: http://jmlr.org/papers/v9/vandermaaten08a.html

[29]   J. Duncombe, "How Machine Learning Redraws the Map of Ocean Ecosystems," *Eos*. http://eos.org/articles/how-machine-learning-redraws-the-map-of-ocean-ecosystems, Jun. 2020.

[30]   M. Sonnewald, S. Dutkiewicz, C. Hill, and G. Forget, "Elucidating Ecological Complexity: Unsupervised Learning Determines Global Marine Eco-provinces," *Science Advances*, vol. 6, no. 22, p. eaay4740, doi: 10.1126/sciadv.aay4740.

[31]   T. R. Dieter and H. Zisgen, "Evaluation of the Explanatory Power Of Layer-wise Relevance Propagation Using Adversarial Examples," *Neural Process Lett*, vol. 55, no. 7, pp. 8531–8550, Dec. 2023, doi: 10.1007/s11063-023-11166-8.

[32]   F. V. Davenport and N. S. Diffenbaugh, "Using Machine Learning to Analyze Physical Causes of Climate Change: A Case Study of U.S. Midwest Extreme Precipitation," *Geophysical Research Letters*, vol. 48, no. 15, p. e2021GL093787, Jul. 2021, doi: 10.1029/2021GL093787.

[33]   M. Rahnemoonfar, M. Yari, J. Paden, L. Koenig, and O. Ibikunle, "Deep Multi-scale Learning for Automatic Tracking of Internal Layers of Ice in Radar Data," *Journal of Glaciology*, vol. 67, no. 261, pp. 39–48, Oct. 2020, doi: 10.1017/jog.2020.80.

[34]   M. Fielding, J. Barrott, and A. Flensburg, "Using Artificial Intelligence to Detect Permafrost Thawing." Apr. 08, 2021. Available: https://www.sei.org/featured/artificial-intelligence-permafrost-thawing/

[35]   J. Atkinson, "Severe storms show off their 'plume-age'." Aug. 15, 2018. Available: https://climate.nasa.gov/news/2782/severe-storms-show-off-their-plume-age

[36]   M. Brandt *et al.*, "An Unexpectedly Large Count of Trees in the West African Sahara and Sahel," *Nature*, vol. 587, no. 7832, pp. 78–82, Nov. 2020, doi: 10.1038/s41586-020-2824-5.

[37]   M. Maskey, "A Collective Agenda for AI on the Earth Sciences (AI for Good)." Feb. 2022. Available: https://www.youtube.com/watch?v=ts5XSYgcsiE

[38]   J. Reinders, "Topology Can Help Us Find Patterns in Weather." Dec. 06, 2018. Available: https://www.hpcwire.com/2018/12/06/topology-can-help-us-find-patterns-in-weather/

[39]   S. Salcedo-Sanz *et al.*, "Machine Learning Information Fusion in Earth Observation: A Comprehensive Review of Methods, Applications and Data Sources," *Information Fusion*, vol. 63, pp. 256–272, Nov. 2020, doi: 10.1016/j.inffus.2020.07.004.

[40]   M. Reichstein *et al.*, "Deep Learning and Process Understanding for Data-driven Earth System Science," *Nature*, vol. 566, no. 7743, 7743, pp. 195–204, Feb. 2019, doi: 10.1038/s41586-019-0912-1.

[41]   T. Schneider *et al.*, "Climate Goals and Computing the Future of Clouds," *Nature Climate Change*, vol. 7, no. 1, pp. 3–5, Jan. 2017, doi: 10.1038/nclimate3190.

[42]   P. Bauer, A. Thorpe, and G. Brunet, "The Quiet Revolution of Numerical Weather Prediction," *Nature*, vol. 525, no. 7567, 7567, pp. 47–55, Sep. 2015, doi: 10.1038/nature14956.

[43]   I. Lopez-Gomez, C. Christopoulos, H. L. Langeland Ervik, O. R. A. Dunbar, Y. Cohen, and T. Schneider, "Training Physics-Based Machine-Learning Parameterizations with Gradient-Free Ensemble Kalman Methods," *Journal of Advances in Modeling Earth Systems*, vol. 14, no. 8, p. e2022MS003105, 2022, doi: 10.1029/2022MS003105.

[44] G. Behrens, T. Beucler, P. Gentine, F. Iglesias-Suarez, M. Pritchard, and V. Eyring, "Non-Linear Dimensionality Reduction With a Variational Encoder Decoder to Understand Convective Processes in Climate Models," *Journal of Advances in Modeling Earth Systems*, vol. 14, no. 8, p. e2022MS003130, 2022, doi: 10.1029/2022MS003130.

[45] R. Lagerquist, A. McGovern, C. R. Homeyer, D. J. G. Ii, and T. Smith, "Deep Learning on Three-Dimensional Multiscale Data for Next-Hour Tornado Prediction," *Monthly Weather Review*, vol. 148, no. 7, pp. 2837–2861, Jun. 2020, doi: 10.1175/MWR-D-19-0372.1.

[46] T. Kurth *et al.*, "FourCastNet: Accelerating Global High-Resolution Weather Forecasting Using Adaptive Fourier Neural Operators." arXiv, Aug. 2022. doi: 10.48550/arXiv.2208.05419.

[47] R. Lam *et al.*, "Learning Skillful Medium-range Global Weather Forecasting," *Science*, vol. 382, no. 6677, pp. 1416–1421, Dec. 2023, doi: 10.1126/science.adi2336.

[48] K. Bi, L. Xie, H. Zhang, X. Chen, X. Gu, and Q. Tian, "Accurate Medium-range Global Weather Forecasting with 3D Neural Networks," *Nature*, vol. 619, no. 7970, 7970, pp. 533–538, Jul. 2023, doi: 10.1038/s41586-023-06185-3.

[49] P. Laloyaux, T. Kurth, P. D. Dueben, and D. Hall, "Deep Learning to Estimate Model Biases in an Operational NWP Assimilation System," *Journal of Advances in Modeling Earth Systems*, vol. 14, no. 6, p. e2022MS003016, 2022, doi: 10.1029/2022MS003016.

[50] K. Kaheman, S. L. Brunton, and J. N. Kutz, "Automatic Differentiation to Simultaneously Identify Nonlinear Dynamics and Extract noise Probability Distributions from Data," *Machine Learning: Science and Technology*, vol. 3, no. 1, p. 015031, Mar. 2022, doi: 10.1088/2632-2153/ac567a.

[51] M. R. Ebers, K. M. Steele, and J. N. Kutz, "Discrepancy Modeling Framework: Learning Missing Physics, Modeling Systematic Residuals, and Disambiguating between Deterministic and Random Effects," *SIAM Journal on Applied Dynamical Systems*, pp. 440–469, Mar. 2024, doi: 10.1137/22M148375X.

[52] M. Sparkes, "Huge Protein Breakthrough," *New Scientist*, vol. 255, no. 3398, pp. 10–11, Aug. 2022, doi: 10.1016/S0262-4079(22)01372-0.

[53]   R. D. Ball *et al.*, "Evidence for Intrinsic Charm Quarks in the Proton," *Nature*, vol. 608, no. 7923, pp. 483–487, Aug. 2022, doi: 10.1038/s41586-022-04998-2.

[54]   G. Camps-Valls, D. Tuia, X. X. Zhu, and M. Reichstein, *Deep Learning for the Earth Sciences: A Comprehensive Approach to Remote Sensing, Climate Science and Geosciences.* Wiley, Hoboken, New Jersey 2021.

[55]   S. Mahajan, L. S. Passarella, F. M. Hoffman, M. G. Meena, and M. Xu, "Assessing Teleconnections-Induced Predictability of Regional Water Cycle on Seasonal to Decadal Timescales Using Machine Learning Approaches," Artificial Intelligence for Earth System Predictability (AI4ESP) Collaboration (United States), AI4ESP-1086, Apr. 2021. doi: 10.2172/1769676.

[56]   Y. Liu, K. Duffy, J. G. Dy, and A. R. Ganguly, "Explainable Deep Learning for Insights in El Niño and River Flows," *Nat Commun*, vol. 14, no. 1, 1, p. 339, Jan. 2023, doi: 10.1038/s41467-023-35968-5.

[57]   A. Mercer, "Predictability of Common Atmospheric Teleconnection Indices Using Machine Learning," *Procedia Computer Science*, vol. 168, pp. 11–18, Jan. 2020, doi: 10.1016/j.procs.2020.02.245.

[58]   K. Dijkstra, J. van de Loosdrecht, L. R. B. Schomaker, and M. A. Wiering, "Hyperspectral Demosaicking and Crosstalk Correction Using Deep Learning," *Machine Vision and Applications*, vol. 30, no. 1, pp. 1–21, Feb. 2019, doi: 10.1007/s00138-018-0965-4.

[59]   Y. Xiong, Y. Ye, H. Zhang, J. He, B. Wang, and K. Yang, "Deep Learning and Hierarchical Graph-assisted Crosstalk-aware Fragmentation Avoidance Strategy in Space Division Multiplexing Elastic Optical Networks," *Optics Express*, vol. 28, no. 3, pp. 2758–2777, Feb. 2020, doi: 10.1364/OE.381551.

[60]   N. Erfanian *et al.*, "Deep Learning Applications in Single-cell Genomics and Transcriptomics Data Analysis," *Biomedicine & Pharmacotherapy*, vol. 165, p. 115077, Sep. 2023, doi: 10.1016/j.biopha.2023.115077.

[61]   N. Papernot, P. McDaniel, I. Goodfellow, S. Jha, Z. B. Celik, and A. Swami, "Practical Black-Box Attacks against Machine Learning," in *Proceedings of the 2017 ACM on Asia Conference on Computer and Communications Security*, Apr. 2017, pp. 506–519. doi: 10.1145/3052973.3053009.

[62] D. George and E. A. Huerta, "Deep Learning for Real-time Gravitational Wave Detection and Parameter Estimation: Results with Advanced LIGO Data," *Physics Letters B*, vol. 778, pp. 64–70, Mar. 2018, doi: 10.1016/j.physletb.2017.12.053.

[63] A. Sinha and R. Abernathey, "Estimating Ocean Surface Currents with Machine Learning," *Frontiers in Marine Science*, vol. 8, 2021.

[64] M. Raissi and G. E. Karniadakis, "Hidden Physics Models: Machine Learning of Nonlinear Partial Differential Equations," *Journal of Computational Physics*, vol. 357, pp. 125–141, Mar. 2018, doi: 10.1016/j.jcp.2017.11.039.

[65] M. Raissi, P. Perdikaris, and G. E. Karniadakis, "Physics-informed Neural Networks: A Deep Learning Framework for Solving Forward and Inverse Problems Involving Nonlinear Partial Differential Equations," *Journal of Computational Physics*, vol. 378, pp. 686–707, Feb. 2019, doi: 10.1016/j.jcp.2018.10.045.

[66] E. P. L. van Nieuwenburg, Y.-H. Liu, and S. D. Huber, "Learning Phase Transitions by Confusion," *Nature Physics*, vol. 13, no. 5, pp. 435–439, Feb. 2017, doi: 10.1038/nphys4037.

[67] S. Srinivasan *et al.*, "Machine Learning the Metastable Phase Diagram of Covalently Bonded Carbon," *Nature Communications*, vol. 13, no. 1, 1, p. 3251, Jun. 2022, doi: 10.1038/s41467-022-30820-8.

[68] A. S. von der Heydt *et al.*, "Lessons on Climate Sensitivity From Past Climate Changes," *Current Climate Change Reports*, vol. 2, no. 4, pp. 148–158, Dec. 2016, doi: 10.1007/s40641-016-0049-3.

[69] N. Wolchover, "Machine Learning's 'Amazing' Ability to Predict Chaos," *Quanta Magazine*. https://www.quantamagazine.org/machine-learnings-amazing-ability-to-predict-chaos-20180418/, Apr. 2018.

[70] J. Pathak, B. Hunt, M. Girvan, Z. Lu, and E. Ott, "Model-Free Prediction of Large Spatiotemporally Chaotic Systems from Data: A Reservoir Computing Approach," *Physical Review Letters*, vol. 120, no. 2, p. 024102, Jan. 2018, doi: 10.1103/PhysRevLett.120.024102.

[71] A. Krizhevsky, I. Sutskever, and G. E. Hinton, "ImageNet Classification with Deep Convolutional Neural Networks," *Communications of the ACM*, vol. 60, no. 6, pp. 84–90, May 2017, doi: 10.1145/3065386.

[72]   Y. LeCun *et al.*, "Backpropagation Applied to Handwritten Zip Code Recognition," *Neural Computation*, vol. 1, no. 4, pp. 541–551, Dec. 1989, doi: 10.1162/neco.1989.1.4.541.

[73]   J. Schmidhuber, "Who Invented Backpropagation?" Nov. 2020. Available: https://people.idsia.ch/~juergen/who-invented-backpropagation.html

[74]   A. Halevy, P. Norvig, and F. Pereira, "The Unreasonable Effectiveness of Data," *IEEE Intelligent Systems*, vol. 24, no. 2, pp. 8–12, Mar. 2009, doi: 10.1109/MIS.2009.36.

[75]   R. Sutton, "The Bitter Lesson," Mar. 19, 2019. http://www.incompleteideas.net/IncIdeas/BitterLesson.html (accessed Feb. 09, 2024).

[76]   J. Kaplan *et al.*, "Scaling Laws for Neural Language Models," Jan. 22, 2020. http://arxiv.org/abs/2001.08361 (accessed Feb. 09, 2024).

[77]   Epoch, "Parameter, Compute and Data Trends in Machine Learning." https://epochai.org/data/epochdb, 2024.

[78]   S. Chetlur *et al.*, "cuDNN: Efficient Primitives for Deep Learning," doi: 10.48550/arXiv.1410.0759.

[79]   T. Hoefler, D. Alistarh, T. Ben-Nun, N. Dryden, and A. Peste, "Sparsity in Deep Learning: Pruning and Growth for Efficient Inference and Training in Neural Networks," *Journal of Machine Learning Research*, vol. 22, no. 241, pp. 1–124, 2021.

[80]   M. Speiser, "On Sparsity in AI/ML and Earth Science Applications, and Its Architectural Implications." Geneva, Switzerland/Virtual, Jul. 2021.

[81]   M. Radosavljevic and J. Kavalieros, "3D-Stacked CMOS Takes Moore's Law to New Heights," *IEEE Spectrum*. https://spectrum.ieee.org/3d-cmos, Aug. 2022.

[82]   Lex Fridman, "Scott Aaronson: Quantum Computing." Feb. 17, 2020. Available: https://www.youtube.com/watch?v=uX5t8Eiv CaM

[83]   F. Tennie and T. N. Palmer, "Quantum Computers for Weather and Climate Prediction: The Good, the Bad, and the Noisy," *Bulletin of the American Meteorological Society*, vol. 104, no. 2, pp. E488–E500, Feb. 2023, doi: 10.1175/BAMS-D-22-0031.1.

[84]   A. Jolly, "Researchers Simulate Ice Formation by Combining AI and Quantum Mechanics," *HPCwire*. https://www.hpcwire.com/off-the-wire/researchers-simulate-ice-formation-by-combining-ai-and-quantum-mechanics/, Aug. 2022.

[85]   L. Zhang, J. Han, H. Wang, R. Car, and W. E, "Deep Potential Molecular Dynamics: A Scalable Model with the Accuracy of Quantum Mechanics," *Physical Review Letters*, vol. 120, no. 14, p. 143001, Apr. 2018, doi: 10.1103/PhysRevLett.120.143001.

[86]   Sabine Hossenfelder, *It Looks Like AI Will Kill Quantum Computing*. 2024. Accessed: Feb. 21, 2024. [Online]. Available: https://www.youtube.com/watch?v=Q8A4wEohqT0

[87]   N. Oreskes, "Why Believe a Computer? Models, Measures, and Meaning in the Natural World," in *The Earth Around Us*, Routledge, 2000.

[88]   N. Oreskes, "The Role of Quantitative Models in Science," in *Models in Ecosystem Science*, C. D. Canham, J. J. Cole, and W. K. Lauenroth, Eds. Princeton University Press, 2003.

[89]   D. Reed *et al.*, "Computational Science: Ensuring America's Competitiveness," p. 117, Jun. 2005.

[90]   T. Hey, S. Tansley, and K. Tolle, *The Fourth Paradigm: Data-Intensive Scientific Discovery*, 1st ed. Microsoft Research, Redmond, Washington 2009.

[91]   A. Calhoun, "How Do Deep Networks of AI Learn?" *Simons Foundation*. https://www.simonsfoundation.org/2020/02/26/how-do-deep-networks-of-ai-learn/, Feb. 2020.

[92]   G. E. Hinton, N. Srivastava, A. Krizhevsky, I. Sutskever, and R. R. Salakhutdinov, "Improving Neural Networks by Preventing Co-adaptation of Feature Detectors," *arXiv:1207.0580 [cs]*, Jul. 2012, Available: https://arxiv.org/abs/1207.0580

[93]   C. Metz, "In Two Moves, AlphaGo and Lee Sedol Redefined the Future," *Wired*, Mar. 2016.

[94]   K. Hartnett, "Machine Learning Confronts the Elephant in the Room," *Quanta Magazine*. https://www.quantamagazine.org/machine-learning-confronts-the-elephant-in-the-room-20180920/, Sep. 2018.

[95]   F. Boutier, "New Copernicus Data Access Service," *European Commission*. https://ec.europa.eu/commission/presscorner/detail/en/ip_22_7374, Dec. 2022.

[96]   "ESDS Program – Continuous Evolution," *NASA Earth-data.* https://www.earthdata.nasa.gov/esds/continuous-evolution; Earth Science Data Systems, NASA, Jan. 2024.

[97]   A. Ng, Y. B. Mourri, and K. Katanforoosh, "Structuring Machine Learning Projects [MOOC]," *Coursera.* https://www.coursera.org/learn/machine-learning-projects, 2021.

[98]   M. G. Schultz *et al.*, "Can Deep Learning Beat Numerical Weather Prediction?" *Philosophical Transactions of the Royal Society A: Mathematical, Physical and Engineering Sciences*, vol. 379, no. 2194, p. 20200097, Apr. 2021, doi: 10.1098/rsta.2020.0097.

[99]   G. Cybenko, "Approximation by Superpositions of a Sigmoidal Function," *Mathematics of Control, Signals and Systems*, vol. 2, no. 4, pp. 303–314, Dec. 1989, doi: 10.1007/BF02551274.

[100] K. Hornik, M. Stinchcombe, and H. White, "Multilayer Feed-forward Networks Are Universal Approximators," *Neural Networks*, vol. 2, no. 5, pp. 359–366, Jan. 1989, doi: 10.1016/0893-6080(89)90020-8.

[101] J. Deng, W. Dong, R. Socher, L.-J. Li, K. Li, and L. Fei-Fei, "ImageNet: A Large-scale Hierarchical Image Database," in *2009 IEEE Conference on Computer Vision and Pattern Recognition*, Jun. 2009, pp. 248–255. doi: 10.1109/CVPR.2009.5206848.

[102] J. Markoff, "Seeking a Better Way to Find Web Images," *The New York Times*, Nov. 2012.

[103] A. Radford, K. Narasimhan, T. Salimans, and I. Sutskever, "Improving Language Understanding by Generative Pre-Training," *OpenAI Blog.* 2018.

[104] A. Radford, J. Wu, R. Child, D. Luan, D. Amodei, and I. Sutskever, "Language Models Are Unsupervised Multitask Learners," *OpenAI Blog.* 2019.

[105] H. Touvron *et al.*, "LLaMA: Open and Efficient Foundation Language Models." arXiv, Feb. 2023. doi: 10.48550/arXiv.2302.13971.

[106] H. Touvron *et al.*, "Llama 2: Open Foundation and Fine-Tuned Chat Models." arXiv, Jul. 2023. doi: 10.48550/arXiv.2307.09288.

[107] J. Banks and T. Warkentin, "Gemma: Introducing new state-of-the-art open models," *Google.* Feb. 2024.

[108] BigScience Workshop *et al.*, "BLOOM: A 176B-Parameter Open-Access Multilingual Language Model." arXiv, Jun. 2023. doi: 10.48550/arXiv.2211.05100.

[109] A. Ramesh *et al.*, "Zero-Shot Text-to-Image Generation." arXiv, Feb. 2021. doi: 10.48550/arXiv.2102.12092.

[110] A. Ramesh, P. Dhariwal, A. Nichol, C. Chu, and M. Chen, "Hierarchical Text-Conditional Image Generation with CLIP Latents." arXiv, Apr. 2022. doi: 10.48550/arXiv.2204.06125.

[111] J. Betker *et al.*, "Improving Image Generation with Better Captions," *OpenAI Blog*. Oct. 2023.

[112] O. Bar-Tal *et al.*, "Lumiere: A Space-Time Diffusion Model for Video Generation," Feb. 05, 2024. http://arxiv.org/abs/2401.129 45 (accessed Feb. 06, 2024).

[113] "Video Generation Models as World Simulators," *OpenAI Blog*. Feb. 2024.

[114] J. Porter, "ChatGPT Continues to Be One of the Fastest-growing Services Ever," *The Verge*. https://www.theverge.com/2023/ 11/6/23948386/chatgpt-active-user-count-openai-developer-conference, Nov. 2023.

[115] J. Nicas and L. C. Herrera, "Is Argentina the First A.I. Election?" *The New York Times*, Nov. 2023.

[116] Z. Wolf, "Analysis: The Deepfake Era of US Politics Is upon Us | CNN Politics," *CNN*. https://www.cnn.com/2024/01/24/politics/ deepfake-politician-biden-what-matters/index.html, Jan. 2024.

[117] J. G. Cavazos, P. J. Phillips, C. D. Castillo, and A. J. O'Toole, "Accuracy Comparison across Face Recognition Algorithms: Where Are We on Measuring Race Bias?" *IEEE Transactions on Biometrics, Behavior, and Identity Science*, vol. 3, no. 1, pp. 101–111, Jan. 2021, doi: 10.1109/TBIOM.2020.3027269.

[118] "Article 5: Prohibited Artificial Intelligence Practices | EU Artificial Intelligence Act (Final Draft January 2024)," *EU Artificial Intelligence Act*. https://artificialintelligenceact.eu/article/5/.

[119] "Article 52: Transparency Obligations for Providers and Users of Certain AI Systems and GPAI Models | EU Artificial Intelligence Act (Final Draft January 2024)," *EU Artificial Intelligence Act*. https://artificialintelligenceact.eu/article/52/.

[120] N. Carlini *et al.*, "Extracting Training Data from Diffusion Models." arXiv, Jan. 2023. doi: 10.48550/arXiv.2301.13188.

[121] M. Nasr *et al.*, "Scalable Extraction of Training Data from (Production) Language Models." arXiv, Nov. 2023. doi: 10.48550/arXiv.2311.17035.

[122] I. Shumailov, Z. Shumaylov, Y. Zhao, Y. Gal, N. Papernot, and R. Anderson, "The Curse of Recursion: Training on Generated Data Makes Models Forget," May 31, 2023. http://arxiv.org/abs/2305.17493 (accessed Feb. 01, 2024).

[123] Y. Guo, G. Shang, M. Vazirgiannis, and C. Clavel, "The Curious Decline of Linguistic Diversity: Training Language Models on Synthetic Text." arXiv, Nov. 2023. doi: 10.48550/arXiv.2311.09807.

[124] R. Hataya, H. Bao, and H. Arai, "Will Large-scale Generative Models Corrupt Future Datasets?" arXiv, Aug. 2023. doi: 10.48550/arXiv.2211.08095.

[125] J. Pathak *et al.*, "FourCastNet: A Global Data-driven High-resolution Weather Model Using Adaptive Fourier Neural Operators," Feb. 22, 2022. http://arxiv.org/abs/2202.11214 (accessed Jan. 30, 2024).

[126] A. J. Charlton-Perez *et al.*, "Do AI Models Produce Better Weather Forecasts Than Physics-based Models? A Quantitative Evaluation Case Study of Storm Ciarán," *npj Climate and Atmospheric Science*, vol. 7, no. 1, pp. 1–11, Apr. 2024, doi: 10.1038/s41612-024-00638-w.

[127] M. J. Smith, L. Fleming, and J. E. Geach, "EarthPT: A Time Series Foundation Model for Earth Observation." arXiv, Jan. 2024. doi: 10.48550/arXiv.2309.07207.

[128] G. Mateo-García, V. Laparra, C. Requena-Mesa, and L. Gómez-Chova, "Generative Adversarial Networks in the Geosciences," in *Deep Learning for the Earth Sciences*, John Wiley & Sons, Ltd, 2021, pp. 24–36. doi: 10.1002/9781119646181.ch3.

[129] Lex Fridman, "Yann Lecun: Meta AI, Open Source, Limits of LLMs, AGI & the Future of AI | Lex Fridman Podcast #416." Mar. 07, 2024. Available: https://www.youtube.com/watch?v=5t1vTLU7s40

[130] M. Assran *et al.*, "Self-Supervised Learning from Images with a Joint-Embedding Predictive Architecture." arXiv, Apr. 2023. doi: 10.48550/arXiv.2301.08243.

[131] T. H. Trinh, Y. Wu, Q. V. Le, H. He, and T. Luong, "Solving Olympiad Geometry without Human Demonstrations," *Nature*, vol. 625, no. 7995, 7995, pp. 476–482, Jan. 2024, doi: 10.1038/s41586-023-06747-5.

[132] M. Zečević, M. Willig, D. S. Dhami, and K. Kersting, "Causal Parrots: Large Language Models May Talk Causality but Are Not Causal," *Transactions on Machine Learning Research*, May 2023.

[133] E. M. Bender, T. Gebru, A. McMillan-Major, and S. Shmitchell, "On the Dangers of Stochastic Parrots: Can Language Models Be Too Big?" in *Proceedings of the 2021 ACM Conference on Fairness, Accountability, and Transparency*, Mar. 2021, pp. 610–623. doi: 10.1145/3442188.3445922.

[134] O. J. Dunn, "Multiple Comparisons among Means," *Journal of the American Statistical Association*, vol. 56, no. 293, pp. 52–64, Mar. 1961, doi: 10.1080/01621459.1961.10482090.

[135] J. Neyman and E. S. Pearson, "On the Use and Interpretation of Certain Test Criteria for Purposes of Statistical Inference: Part I," *Biometrika*, vol. 20A, no. 1/2, pp. 175–240, 1928, doi: 10.2307/2331945.

[136] J. Pearl, *Causality*, 2nd ed. Cambridge: Cambridge University Press, 2009. doi: 10.1017/CBO9780511803161.

[137] D. B. Rubin, "Causal Inference Using Potential Outcomes: Design, Modeling, Decisions," *Journal of the American Statistical Association*, vol. 100, no. 469, pp. 322–331, 2005, Accessed: Mar. 28, 2024. [Online]. Available: https://www.jstor.org/stable/27590541

[138] P. Hoyer, D. Janzing, J. M. Mooij, J. Peters, and B. Schölkopf, "Nonlinear Causal Discovery with Additive Noise Models," in *Advances in Neural Information Processing Systems*, 2008, vol. 21.

[139] S. Shimizu, P. O. Hoyer, A. Hyvärinen, and A. Kerminen, "A Linear Non-Gaussian Acyclic Model for Causal Discovery," *Journal of Machine Learning Research*, vol. 7, no. 72, pp. 2003–2030, 2006.

[140] J. Runge, P. Nowack, M. Kretschmer, S. Flaxman, and D. Sejdinovic, "Detecting and Quantifying Causal Associations in Large Nonlinear Time Series Datasets," *Science Advances*, Nov. 2019, doi: 10.1126/sciadv.aau4996.

[141] A. Molak, *Causal Inference and Discovery in Python: Unlock the Secrets of Modern Causal Machine Learning with DoWhy, EconML, PyTorch and More.* Birmingham: Packt Publishing, 2023.

[142] D. Kahneman, *Thinking, Fast and Slow*. New York: Farrar, Straus and Giroux, 2011.

[143] V. Chernozhukov *et al.*, "Double/debiased Machine Learning for Treatment and Structural Parameters," *The Econometrics Journal*, vol. 21, no. 1, pp. C1–C68, Feb. 2018, doi: 10.1111/ectj.12097.

[144]   B. K. Petersen, M. L. Larma, T. N. Mundhenk, C. P. Santiago, S. K. Kim, and J. T. Kim, "Deep Symbolic Regression: Recovering Mathematical Expressions from Data via Risk-seeking Policy Gradients," Feb. 2022.

[145]   W. Tenachi, R. Ibata, and F. I. Diakogiannis, "Deep Symbolic Regression for Physics Guided by Units Constraints: Toward the Automated Discovery of Physical Laws," *The Astrophysical Journal*, vol. 959, no. 2, p. 99, Dec. 2023, doi: 10.3847/1538-4357/ad014c.

[146]   G. Camps-Valls *et al.*, "Discovering Causal Relations and Equations from Data," *Physics Reports*, vol. 1044, pp. 1–68, Dec. 2023, doi: 10.1016/j.physrep.2023.10.005.

[147]   P. D. Dueben, M. G. Schultz, M. Chantry, D. J. Gagne, D. M. Hall, and A. McGovern, "Challenges and Benchmark Datasets for Machine Learning in the Atmospheric Sciences: Definition, Status, and Outlook," *Artificial Intelligence for the Earth Systems*, vol. 1, no. 3, Jul. 2022, doi: 10.1175/AIES-D-21-0002.1.

[148]   J. Kaltenborn *et al.*, "ClimateSet: A Large-Scale Climate Model Dataset for Machine Learning," *Advances in Neural Information Processing Systems*, vol. 36, pp. 21757–21792, Dec. 2023.

[149]   D. Carlson and T. Oda, "Editorial: Data Publication – ESSD Goals, Practices and Recommendations," *Earth System Science Data*, vol. 10, no. 4, pp. 2275–2278, Dec. 2018, doi: 10.5194/essd-10-2275-2018.

[150]   Z. Xiong, F. Zhang, Y. Wang, Y. Shi, and X. X. Zhu, "EarthNets: Empowering AI in Earth Observation." arXiv, Dec. 2022. doi: 10.48550/arXiv.2210.04936.

[151]   S. Rasp *et al.*, "WeatherBench 2: A Benchmark for the Next Generation of Data-driven Global Weather Models," Jan. 26, 2024. http://arxiv.org/abs/2308.15560 (accessed Feb. 20, 2024).

Index

acyclic graph, 67
AGI, *see* artificial general
    intelligence (AGI)
AlexNet, 35, 36, 38, 46, 51
AlphaFold, 26, 38
AlphaGeometry, 60
AlphaGo, 16, 38, 47
architecture
    computer hardware, 2, 37,
    41, 43
    neural network, 12, 13, 15,
    25, 53, 54
Arctic Oscillation, 29, 69
artificial general intelligence
    (AGI), 71
autograd, 8

backdoor adjustment, 66
bias
    bias term, 10
    data bias, 23
    model bias, 23, 25, 26, 56,
    57, 69
    selection bias, 61
BLOOM, 53

causal diagram, 63
causal discovery, 67–69
causal graph, 63–67
causal inference, 65, 66
causal model, 18
causal sufficiency, 67

causation, 28, 61
chaos, 33
ChatGPT, 3, 40, 53, 55, 58, *see*
    generative pre-trained
    transformer (GPT)
CNN, *see* convolutional neural
    network
confabulation, 55
confounder, 61, 66
confounding, 69
convolutional neural network
    (CNN), 12–15, 21, 51
copyright, 71
correlation, 18, 28, 61
    spurious correlation, 62, 63
cross-talk, 29
CUDA, 37
CUDA Deep Neural Network
    library (cuDNN), 37, *see also*
    CUDA
cuDNN, *see* CUDA Deep Neural
    Network library (cuDNN)

DALL-E, 54
deep learning
    definition, 10
    history, 51
deepfake, 56, 57
DeepMind, 16, 25, 26, 53, 60
diffusion model, 17, 54
do-calculus, 65
do-operator, 65

dropout, 12, 46

ECMWF, *see* European centre for medium-range weather forecasts (ECMWF)
El Niño Southern Oscillation (ENSO), 28
ENSO, *see* El Niño Southern Oscillation (ENSO)
ERA5, *see* European centre for medium-range weather forecasts (ECMWF)
European centre for medium-range weather forecasts (ECMWF), 25
ECMWF Reanalysis v5 (ERA5), 26, 58

faithfulness assumption, 67
feature engineering, 11
feedback loop, 31, 67
fine-tuning, 18, 26, 52, 56
FourCastNet, 25, 58

GenAI, *see* generative model
generative AI, *see* generative model
generative model, 17, 53, 54, 58
generative pre-trained transformer (GPT), 53, 54
GNN, *see* graph neural network (GNN)
Google, 36, 38, 54, 62, 63
GPT, *see* generative pre-trained transformer (GPT)
GPU, *see* graphics processing unit (GPU)
graph neural network (GNN), 25, 58
GraphCast, 25

graphics processing unit (GPU), 11, 15, 35, 37, 51

hallucination, 55
high-performance computing (HPC), 2
HPC, *see* high-performance computing (HPC)

ILSVRC, *see* ImageNet large-scale visual recognition challenge (ILSVRC)
ImageNet large-scale visual recognition challenge (ILSVRC), 36, 38, 46, 51, 52
intervention, 65, 66, 68

large language model (LLM), 3, 6, 16, 37, 40, 53, 55, 59, 60
latent diffusion, 18, *see also* diffusion model
layer-wise relevance propagation (LRP), 22
Llama, 53
LLM, *see* large language model (LLM)
long short-term memory (LSTM), 38, *see also* recurrent neural network
loss, 8, 9
LRP, *see* layer-wise relevance propagation (LRP)
LSTM, *see* long short-term memory (LSTM)
Lumiere, 54

Markov assumption, 67
model collapse, 58
Moore's law, 37, 41

numerical weather prediction (NWP), 58, 59

NWP, *see* numerical weather prediction (NWP)

OpenAI, 53, 54
overfitting, 8

Pangu-Weather, 25, 58
PCMCI algorithm, 68
phase changes, 32
potential outcomes, 63

rectified linear unit (ReLU), 10
recurrent neural network (RNN), 15
regularization, 9
reinforcement learning (RL), 16, 70
reinforcement learning from human feedback (RLHF), 16, 53, *see also* reinforcement learning
ReLU, *see* rectified linear unit (ReLU)
RL, *see* reinforcement learning
RLHF, *see* reinforcement learning from human feedback (RLHF)
RNN, *see* recurrent neural network

saliency map, 21, 28
selection bias, *see* bias
self-supervised learning, 6

sigmoid, 10
sparsity, 29, 39
stochastic gradient descent (SGD), 7, 8, 46
structural causal model (SCM), 63
symbolic regression, 70

tanh, 10
teleconnection, 28
tensor processing unit (TPU), 38
TensorFlow, 38
test set, 7
text-to-image, 57
TPU, *see* tensor processing unit (TPU)
training data, 7
transfer learning, 18, 52
transformer, 15, 25, 53, *see also* vision transformer (ViT)

U-Net, 54
unsupervised learning, 53

validation set, 7
vision transformer (ViT), 15
ViT, *see* vision transformer (ViT)

Walker circulation, 68
weak signals, 29

Printed in the United States
by Baker & Taylor Publisher Services